SpringerBriefs in Astronomy

Series Editors

Martin Ratcliffe
Valley Center, KS, USA

Wolfgang Hillebrandt
MPI für Astrophysik, Garching, Germany

Michael Inglis
Department of Physical Sciences, SUNY Suffolk County Community College,
Selden, NY, USA

David Weintraub
Department of Physics & Astronomy, Vanderbilt University, Nashville, TN, USA

SpringerBriefs in Astronomy are a series of slim high-quality publications encompassing the entire spectrum of Astronomy, Astrophysics, Astrophysical Cosmology, Planetary and Space Science, Astrobiology as well as History of Astronomy. Manuscripts for SpringerBriefs in Astronomy will be evaluated by Springer and by members of the Editorial Board. Proposals and other communication should be sent to your Publishing Editors at Springer.

Featuring compact volumes of 50 to 125 pages (approximately 20,000–45,000 words), Briefs are shorter than a conventional book but longer than a journal article. Thus Briefs serve as timely, concise tools for students, researchers, and professionals.

- Typical texts for publication might include:
- A snapshot review of the current state of a hot or emerging field
- A concise introduction to core concepts that students must understand in order to make independent contributions
- An extended research report giving more details and discussion than is possible in a conventional journal article
- A manual describing underlying principles and best practices for an experimental technique
- An essay exploring new ideas within astronomy and related areas, or broader topics such as science and society

Briefs allow authors to present their ideas and readers to absorb them with minimal time investment.

Briefs will be published as part of Springer's eBook collection, with millions of readers worldwide. In addition, they will be available, just like other books, for individual print and electronic purchase.

Briefs are characterized by fast, global electronic dissemination, straightforward publishing agreements, easy-to-use manuscript preparation and formatting guidelines, and expedited production schedules. We aim for publication 8–12 weeks after acceptance.

More information about this series at http://www.springer.com/series/10090

Henri M. J. Boffin · David Jones

The Importance of Binaries in the Formation and Evolution of Planetary Nebulae

 Springer

Henri M. J. Boffin
ESO
Garching bei München, Germany

David Jones
IAC
La Palma, Spain

ISSN 2191-9100 ISSN 2191-9119 (electronic)
SpringerBriefs in Astronomy
ISBN 978-3-030-25058-4 ISBN 978-3-030-25059-1 (eBook)
https://doi.org/10.1007/978-3-030-25059-1

This Springer imprint is published by the registered company Springer Nature Switzerland AG
The registered company address is: Gewerbestrasse 11, 6330 Cham, Switzerland

To our families, without whom none of this would have been possible.

Cathy, Yuki and Niall
Flori and Idris

Foreword

The evolution of stars in gravitationally bound binary systems is an extraordinarily rich field of study. Not only can the evolution of each star be dramatically altered with respect to the evolution of single stars, but a number of physical phenomena also occur that are highly relevant to other fields of astrophysics. The very long list includes, but is most certainly not limited to: tidal interactions, surface irradiation, mass transfer and accretion, the production of stellar jets and other outflow phenomena, gravitational wave emission and other relativistic effects. As many of these effects are the stellar analogues of similar phenomena occurring at much larger scales, such as in the cores of galaxies, modern astrophysics cannot be conceived without a thorough understanding of binary stars. Their relevance is further enhanced by the fact that it is believed that more than 50% of stars are found in binary or higher-order systems, which makes binarity a necessary ingredient even for the interpretation of the most basic relationships such as the classical Hertzsprung–Russell diagram.

Binary interactions occur among all types of stars and span a large range of orbital separations, from the shortest binary systems known nowadays, pairs of compact white dwarfs with orbital periods as short as five minutes (HM Cancri), through to giant stars that can be affected by the presence of companions even at orbital periods of many hundreds of years (o Ceti [Mira]).

This book by Astrophysicists Henri M. J. Boffin and David Jones focusses on the effects of binarity in the final evolutionary stages of low- and intermediate-mass stars. The relevance of binary interactions to explain the properties of planetary nebulae has been an active subject of debate since the first high-quality catalogues of narrowband images of planetary nebulae were obtained at the beginning of the 1990s both from the ground and with the Hubble Space Telescope. Today, far fewer astronomers doubt that the key to understanding the wide variety of planetary nebulae shapes, or indeed some of their peculiar chemical properties, is binary evolution. However, much work has still to be done in order to constrain the overall statistical relevance of binarity in the formation and evolution of planetary nebulae as well as the specific physical processes involved and how they are related to observed nebular properties. The topic also has important implications for our understanding of cataclysmic variables and novae, Type Ia supernovae, symbiotic stars and other phenomena such as the production of astrophysical jets.

This book comprehensively outlines current understanding in sufficient detail as to make it a valuable reference text, providing not only a global view of the subject but also guidance for planning the future research in a field that has shown tremendous, albeit still insufficient, progress over the last three decades. I have no doubt that this book will occupy a permanent place on my desk for years to come.

La Palma, Spain   Romano L. M. Corradi

Preface

It is now clear that a binary evolutionary pathway is responsible for a significant fraction of all planetary nebulae (PNe), with some authors even going as far as claiming that the Sun will not become a PN. At the very least, it is now clear that binary interactions play a critical role in the shaping of many PNe—including some of the most well studied. Furthermore, PNe offer a unique window into many key aspects of binary evolution, providing multiple avenues to explore the various physical processes involved. Beyond the central stars themselves, the surrounding nebulae offer an additional route to trace the mass loss and mass transfer histories of these systems, meaning that one can, in principle, derive a complete picture of the impact of binary evolution on these systems. Furthermore, binary central stars of PNe represent progenitor systems for a wide range of astrophysical phenomena, including cosmologically important Type Ia supernovae and the stellar-mass gravitational wave sources that will be revealed by next-generation detectors. This combined with the fact that the majority of stars are found to reside in binary systems, many of which will interact at some point during their lives, only serves to further highlight the importance of understanding the impact of binarity on stellar evolution including the late stages which in intermediate-mass stars are characterised by the formation of a PN. Collectively, the weight of recent advances has led to the requirement that textbooks need to be rewritten. This *SpringerBriefs* is the very first step in this direction.

We have tried, and by no means claim to have succeeded, to present in a succinct way all the theoretical and observational support for the importance of binarity in the formation of PNe. In the process, we outline some of the key principles and techniques, as well as their flaws and advantages, used in the study of binary PNe (many of which have wider applications). As such, we hope that this book will be useful for all specialists, from graduate students to senior astronomers, working in (binary) stellar physics, but also to anyone that is interested in this very important and aesthetically beautiful phase of stellar evolution.

It is a pleasure to thank our many long-suffering collaborators who have contributed significantly to much of the work detailed within this book, including Romano Corradi, Jorge García-Rojas, Alain Jorissen, Dimitri Pourbaix, Pablo

Rodríguez-Gil, Miguel Santander-García, Roger Wesson, Brent Miszalski, Paulina Sowicka, Todd Hillwig, Hans Van Winckel, Alex Brown, Ana Escorza, Alba Aller, Myfanwy Lloyd, Lionel Siess, Sophie Van Eck, the Phoebe development team and many, many more. We also owe a debt of gratitude to the excellent scientists who over the years have contributed enormously to our understanding of planetary nebulae and binary evolution, and whose work serves as the bedrock for this book. HMJB reserves special thanks for Prof. A. Acker for introducing him to this very exciting research field.

Finally, it is important to thank the various funding agencies that have supported much of the work that this book comprises. The initial outline was drafted while HMJB was visiting the IAC, thanks to a visitor grant in the framework of a Severo Ochoa Excellence programme (SEV-2015-0548). DJ gratefully acknowledges support from the State Research Agency (AEI) of the Spanish Ministry of Science, Innovation and Universities (MCIU) and the European Regional Development Fund (FEDER) under grant AYA2017-83383-P. DJ also acknowledges support under grant P/308614 financed by funds transferred from the Spanish Ministry of Science, Innovation and Universities, charged to the General State Budgets and with funds transferred from the General Budgets of the Autonomous Community of the Canary Islands by the Ministry of Economy, Industry, Trade and Knowledge.

Garching bei München, Germany   Henri M. J. Boffin
La Palma, Spain   David Jones

Contents

Acronyms

$M_\odot$	Solar mass
$R_\odot$	Solar radius
ADF	Abundance discrepancy factor
AGB	Asymptotic giant branch
bCSPNe	Binary central stars of planetary nebulae
BRET	Bipolar, rotating, episodic jet
CE	Common envelope
CS	Central star
CSPNe	Central stars of planetary nebulae
CV	Cataclysmic variable
DD	Double degenerate
ESO	European Southern Observatory
FDU	First dredge-up
GE	Grazing envelope
GISW	Generalised interacting stellar wind
HBB	Hot bottom burning
IFMR	Initial–final mass relation
ISW	Interacting stellar wind
LMXB	Low-mass X-ray binary
LSST	Large Synoptic Survey Telescope
MACHO	Massive Compact Halo Object project
MS	Main sequence
NTT	New Technology Telescope
OGLE	Optical Gravitational Lensing Experiment
PN(e)	Planetary nebula(e)
PNLF	Planetary nebula luminosity function
RGB	Red-giant branch
RLOF	Roche lobe overflow
SDU	Second dredge-up
SNe	Supernovae

TDU	Third dredge-up
TP-AGB	Thermally pulsing asymptotic giant branch
VLT	Very Large Telescope
WD	White dwarf
WRLOF	Wind Roche lobe overflow

Chapter 1
Introduction

1.1 Historical Overview

The first planetary nebula (PN) discovered was the Dumbbell nebula by Charles Messier in 1764 (a modern CCD image of which is presented in Fig. 1.1), and to which he assigned the number 27 in his later catalogue of nebulous objects first published in 1774 (the final version of his catalogue included three more PNe; Messier 1781). At the time, the nature of such nebulae was unclear with the term nebula being used as a catch-all for any seemingly diffuse, non-star-like object. It was only from observations with improved resolution that the gaseous nebulae, like PNe, were distinguished from the nebulae made up of stars (like galaxies and star clusters). The term, or rather misnomer, planetary nebula was coined by William Herschel (then court astronomer to King George III), owing to their apparent resemblance to his recent discovery, the planet Uranus (Herschel 1791). It was also Herschel who first began to decipher their nature, with the discovery of bright points at their centres (the central stars), stating in his seminal work "On Nebulous Stars, Properly So Called" (Herschel 1791):

> The nature of planetary nebulae, which has hitherto been involved in much darkness, may now be explained with some degree of satisfaction, since the uniform and very considerable brightness of their apparent disk accords remarkably well with a much condensed, luminous fluid; whereas to suppose them to consist of clustering stars will not so completely account for the milkiness or soft tint of their light, to produce which it would be required for that condensation of the stars should carried to an almost inconceivable degree of accumulation.

Such an interpretation became yet clearer with the first spectroscopic observations of a PN, which showed no continuum emission but rather a series of emission lines. Huggins and Miller (1864) identified three emission lines in their spectrum of NGC 6543 (an image of the nebula is presented in Fig. 1.2 while the original spectrum is shown in Fig. 1.3), the two brightest of which they struggled to associate with a chemical species (they considered Nitrogen but found the absence of nearby expected lines troubling), while the faintest of which was correctly identified as the Fraunhofer F

© The Author(s), under exclusive license to Springer Nature Switzerland AG 2019
H. M. J. Boffin and D. Jones. *The Importance of Binaries in the Formation
and Evolution of Planetary Nebulae*, SpringerBriefs in Astronomy,
https://doi.org/10.1007/978-3-030-25059-1_1

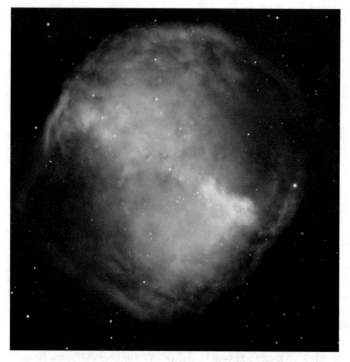

Fig. 1.1 A colour-composite image of the very first PN discovered, M 27, also known as the Dumbbell nebula (Credit: ESO/I. Appenzeller, W. Seifert, O. Stahl)

line (now more commonly known as Hβ). A modern CCD spectrum is shown in Fig. 1.4 with the lines observed by Huggins and Miller (1864) correctly identified. Given that such a spectrum is so removed from the observed spectra of stars (which show continuum emission with absorption features), it was clear that PNe could not be the result of reflected starlight. However, it was not until much later and the work of Zanstra (1927) that it became clear that the emission spectrum of PNe was due to the photoionisation of the nebular material by the hot central star. While this interpretation explained the mechanism by which the multiple lines of Hydrogen and Helium are emitted, it could not provide an explanation for the still unidentified bright lines observed by Huggins and Miller (1864).

It had been suggested that the unidentified lines in the PN spectrum might originate from an as yet undiscovered element dubbed "nebulium" (Nicholson 1911)—the existence of which was all but ruled out by Dimitri Mendeleev's development of the periodic table and Henry Moseley's work on the relationship between the frequency of emitted X-rays and the atomic number of the emitting species (Moseley's law). It was not until Bowen (1927), that it was suggested that the observed lines could perhaps arise from terrestrial chemical species observed in conditions very different to laboratory conditions on Earth (identifying low densities as the likely culprit). Bowen (1928) went on to identify various lines as being collisionally excited lines of singly ionised Nitrogen, as well as singly and doubly ionised Oxygen

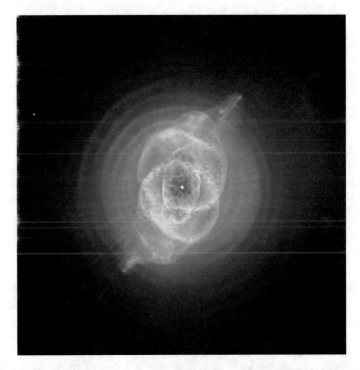

Fig. 1.2 A Hubble Space Telescope colour-composite image of the PN NGC 6543 (Credit: NASA, ESA, HEIC, and The Hubble Heritage Team [STScI/AURA])—the first PN to be observed spectroscopically (Huggins and Miller 1864)

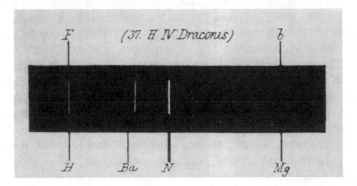

Fig. 1.3 The first spectrum of NGC 6543 (reproduced from Huggins and Miller 1864), including the misidentification of [O III] 5007Å as a line of Nitrogen

(including the lines first observed by Huggins and Miller 1864, which were shown to be [O III] 4959Å and [O III] 5007Å as highlighted in Fig. 1.4). While this essentially solved the question of their excitation mechanism and the origins of their spectra, there was still almost nothing known about the evolutionary origins of PNe.

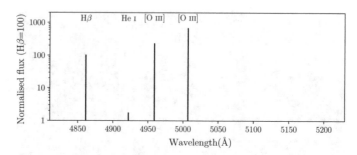

Fig. 1.4 A modern CCD spectrum of NGC 6543 (data originally presented in Middlemass et al. 1989), covering roughly the same spectral range of first observations by Huggins and Miller (1864, shown in Fig. 1.3). Note the logarithmic flux scale highlighting the presence of the much fainter line of He I, which was not detected by Huggins and Miller (1864)

Shklovsky (1956) was the first to suggest that PNe may represent the direct progenitors of white dwarfs, while Abell and Goldreich (1966) argued that PNe are the ejected envelopes of red giant stars, thus establishing the position of PNe in the standard picture of stellar evolution (which will be discussed in greater detail in Sect. 1.2). However, it was not until Kwok et al. (1978) that a satisfactory model was found to explain their physical structures, with most previous authors favouring a form of sudden ejection (dynamical instabilities due to recombination, e.g. Lucy 1967; pulsational instabilities, e.g. Kutter and Sparks 1974; thermal pulses, e.g. Harm and Schwarzschild 1975; and a range of others). These sudden ejection models are all incapable, in one way or another, of reproducing the observed properties of PNe—be that the observed expansion velocities, shell densities and masses, not leaving behind a core hot enough to ionise the nebula before it disperses, or preventing fallback/backfill towards the core. Kwok et al. (1978) proposed an elegant solution, combining the discovery of extensive mass loss from red giants (Gehrz and Woolf 1971) with the evidence for high-velocity winds originating from the exposed cores of PN nuclei, in which the interaction between these two wind phases forms the nebular shell. Under this scheme, known as the Interacting Stellar Winds (ISW) model, the red giant mass loss (in the form of a slow but rather dense stellar wind) simply continues until the entire envelope is exhausted and the hot core of the star is exposed. The faster but more tenuous wind that is then launched from the hot nucleus interacts with the previous mass loss, sweeping it up into a shell which is then ionised by the radiation from the central star. This simple model does not suffer from the aforementioned problems which plagued sudden ejection models and, with the incorporation of asphericities in the winds (known as the Generalised ISW or GISW model; Kahn and West 1985), can even begin to explain the variation in observed PN morphologies (which will be discussed in detail in Sect. 1.3.3).

1.2 The Evolution of Low- and Intermediate-Mass Stars

As PNe represent the final evolutionary stages of their progenitor stars, it is important to quickly summarise the evolution of these stars and the important aspects which will then impact upon the properties of the resulting PNe.

PNe are thought to originate from stars whose initial mass lies in the range $\sim$0.7– 8 $M_\odot$ with lower mass stars evolving too slowly to have yet reached the PN phase while more massive stars are thought to undergo core-collapse at the end of their lives leading to type II supernovae, neutron stars and/or black holes (Pejcha and Thompson 2015). The lives of low- ($M \lesssim 2.3\ M_\odot$) and intermediate-mass ($2.3 \lesssim M \lesssim 8\ M_\odot$) stars are generally very similar, following comparable trajectories across the Herzsprung–Russel (HR) diagram (the evolutionary tracks of typical low- and intermediate-mass stars are illustrated in Fig. 1.5 and Fig. 1.6, respectively). In this section, we will try to briefly cover some of the key aspects of low- and intermediate-mass stellar evolution, from the main sequence through to the end of the asymptotic giant branch (for a much more complete and detailed explanation we refer the reader to Eldridge and Tout 2019).

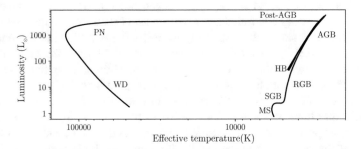

Fig. 1.5 The evolutionary track of a 1 $M_\odot$ star from the main sequence (MS) to the white dwarf (WD) cooling track, calculated using the Modules for Experiments in Stellar Astrophysics code (MESA; Paxton et al. 2011)

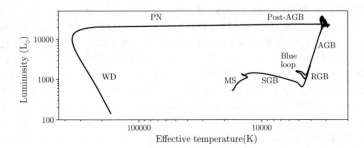

Fig. 1.6 The evolutionary track of a 5 $M_\odot$ star from the main sequence (MS) to the white dwarf (WD) cooling track, calculated using the Modules for Experiments in Stellar Astrophysics code (MESA; Paxton et al. 2011). Note the signature of thermal pulses at the tip of the AGB which are even more prominent than in the 1 $M_\odot$ track (shown in Fig. 1.5) due to the extreme mass-loss experienced by the star at this evolutionary phase

Both low- and intermediate-mass stars burn Hydrogen (H) to Helium (He) in their cores while on the main sequence (with the pp-chain dominating in low-mass stars, while the CNO cycle dominates in the intermediate-mass range). Once the H in the core is exhausted, H-burning continues in a shell further increasing the amount of He ash deposited in the core. As the H-burning shell moves outwards, the stellar envelope expands and cools slightly while maintaining the same luminosity—the star is now on the Sub-Giant Branch (SGB). Stars with masses greater than 1 $M_\odot$ may experience a "hook" on the HR diagram when leaving the main sequence (MS; see Fig. 1.6) due to the delay between the H-burning core being extinguished and an H-burning shell being ignited during which the stars contracts and heats up.

For low-mass stars, as the H-burning shell continues to produce He, the core mass increases until eventually becoming degenerate. At this point its temperature increases, also increasing the burning rate in the H-burning shell. The outer layers of the star become strongly convective and the star increases in luminosity at more or less constant effective temperature—the star is now on the Red Giant Branch (RGB). While on the RGB, convective mixing leads to products of H-fusion being brought to the surface (reducing the C/N and $^{12}C/^{13}C$ ratios) as part of the First Dredge-Up (FDU) episode. At the tip of the RGB, the core temperature is sufficient to ignite He via the triple-α process. As the He core is degenerate prior to ignition, the process is not thermostatically controlled and results in a so-called Helium flash—the brief (a few seconds) runaway that massively increases the star's luminosity until core degeneracy is lifted and stable core He-burning sets in. Shortly afterwards, the He-burning settles down within the convective core, and H-burning continues in a shell just outside the core with the star now on the Horizontal Branch (HB).

In intermediate-mass stars, the He core has already reached the Schönberg-Chandrasekhar limit by the time the star leaves the main sequence, meaning that it is supported primarily by gas pressure (Schönberg and Chandrasekhar 1942). This means that the core is not in thermal equilibrium with the H-burning shell, so it begins to contract while the outer layers of the star expand and cool rather rapidly. As the envelope cools, its opacity increases until it eventually becomes convective and the expansion stops while the stellar luminosity increases (now having reached the RGB just as in low-mass stars). The phase of core contraction and envelope cooling/expansion is extremely short-lived meaning that relatively few intermediate-mass stars are observed in this sub-giant evolutionary phase, earning the region of the HR diagram where these stars lie the name "Hertzsprung gap". While on the RGB, intermediate-mass stars evolve much like low-mass stars, however the He ignition at the RGB tip is rather more gentle. As their cores are above the Schönberg-Chandrasekhar limit, the ignition is thermostatically controlled and does not result in the same runaway Helium flash observed in low-mass stars. Instead, the smooth ignition of He means that the star experiences a "blue loop" (see Fig. 1.6), a name derived from its shape on the HR diagram as the star loops towards higher temperatures before cooling again while increasing in luminosity.

Once the He in the core is exhausted in both low- and intermediate-mass stars, He burning moves to a shell with the core becoming increasingly degenerate as more and more Carbon and Oxygen ash is added—the star is now on the Asymptotic Giant

Branch (AGB). In intermediate-mass stars, the increasing temperature of the core leads to an increase in the He-burning rate causing the layers outside of the He-burning shell to expand and cool, extinguishing the H-burning shell and allowing a Second Dredge-Up (SDU) in the convective envelope to bring CNO products to the stellar surface (increasing the surface abundance of ^{4}He and ^{14}N, while reducing the relative abundances of ^{12}C and ^{16}O). This SDU ends when convection has brought enough fresh H down towards the He shell such that the outer H-burning shell can reignite. The extreme temperature dependence of the triple-α reaction means that the He-burning shell is rather small and, as such, subject to Härm-Schwarzschild instabilities (Schwarzschild and Härm 1965). This leads to the He-burning shell burning faster than the outer H-burning shell can supply it with fuel such that it cannot burn continuously, resulting in thermal pulses (thermally-pulsing AGB or TP-AGB).

Low-mass stars do not experience a SDU episode at the end of core He-burning, with the He-burning shell simply approaching the outer H-burning shell to form a thin inter-shell region which drives thermal pulses in the same way as intermediate-mass stars. In both low- and intermediate-mass stars, thermal pulses drive a Third Dredge-Up (TDU) episode whereby at the end of each pulse (while the H-burning shell is extinguished and the He-burning shell is cooling) the surface convective envelope reaches deep into the inter-shell region. This brings the products of He-burning to the stellar surface, increasing the surface abundances of both ^{12}C and s-process elements. The increased C/O ratio at the stellar surface leads to the star becoming a Carbon star (with almost all O in the atmosphere locked away in the form of CO, such that none is available to form the characteristic oxides found in the spectra of O-rich stars). Meanwhile, C-rich molecules like CN begin to dominate the observed spectrum.

In more massive intermediate-mass stars (M $\gtrsim$ 4 M$_\odot$), hot bottom burning (HBB) converts ^{12}C to ^{14}N in the convective envelope. As a result, the star can return to once again have an O-rich envelope, providing the HBB rate is faster than the rate at which C is brought to the surface by dredge up.

While on the TP-AGB, low- and intermediate-mass stars expel their outer envelopes in the form of a slow, dense stellar wind with mass-loss rates reaching 10^{-5}–10^{-4} M$_\odot$ yr^{-1} due to a complex interplay between dust formation due to shocks from pulsations and radiation pressure onto the dust grains (Lagadec and Zijlstra 2008; McDonald et al. 2018). Once the envelope is exhausted by this dense wind, the star departs the AGB with the now exposed core contracting and heating at almost constant luminosity—the post-AGB phase (van Winckel 2003). While the mass-loss rate has dropped dramatically, the post-AGB wind is much faster than the previous dense wind and it is the interaction between these two winds that leads to the formation of the PN as explained by the GISW theory described in Sect. 1.1. As the PN expands and ultimately dissipates, the central post-AGB star continues to evolve eventually joining the white dwarf (WD) cooling track (Renedo et al. 2010).

Given the clear variations in envelope chemistry depending on the mass of the progenitor, the observed abundance patterns in PNe can be used to shed light on the nature of their progenitors (assuming binary interaction does not cut short the evolution; Jones et al. 2014). For example, PNe with abundance patterns consistent with this HBB (C/O < 1, N/O > 0.5, He/H > 0.125) are known as Type I PNe (Peimbert and Torres-Peimbert 1983; Kingsburgh and Barlow 1994) and are generally thought to have originated from particularly massive progenitors.

1.3 Why Are Planetary Nebulae Important?

Beyond their beauty, PNe are also of interest for a wide variety of other reasons, some of which we will outline in this section.

1.3.1 Chemical Enrichment of the ISM

Given that the vast majority of all stars are of low to intermediate mass, the mass lost on the AGB (and as part of the PN phase) is an important route towards the enrichment of the interstellar medium (ISM). Indeed, these dominate the return of He, C, N and s-process elements to the ISM. Not only does this make PNe important for the formation of future generations of stars and planets, it also means that they are important tracers of the evolution of galaxies. Their intrinsic brightness and the fact that they are found in all types and ages of galaxy means that they can be used to probe the spatially-resolved chemical abundances of galaxies. This becomes particularly important in galaxies which do not present extensive star formation and thus have few or no HII regions which can be used for the same purpose. As described in Sect. 1.2, the nucleosynthesis processes present in AGB stars do not affect the overall O abundance in their envelopes (rather its abundance relative to other species like C, during the TDU, and N, via HBB of C). This means that PNe can be used to probe the initial O abundances when their progenitors formed (which can be used as a direct proxy for metallicity), while the relative abundance O/N can be used to probe the progenitors' masses and ages (Magrini et al. 2009). Collectively this information can be used to study the formation and evolution of the host galaxy (Rahimi et al. 2011; Corradi et al. 2015b).

1.3.2 The Kinematics and Distances of Galaxies

As PNe are bright emission line objects with a significant fraction of this luminosity emitted in the [O III] 5007Å line (up to ~12% of the central star's total luminosity; Gesicki et al. 2018), this line can be observed at reasonable signal to noise out to several Mpc. Furthermore, given the relatively low expansion velocities of PNe (a few tens of km s^{-1}; Weinberger 1989), radial velocity measurements of this line can be used to probe the kinematics of host galaxies. Such studies benefit from the ubiquitous nature of PNe, and use multi-object spectroscopy, Fabry-Perot interferometry or Counter-dispersed spectroscopy of tens or hundreds of PNe at once (Douglas et al. 2002) to derive the rotation curves of entire galaxies in one shot. Here, studies employing PNe are particularly critical for probing dark matter haloes as the stellar density in these regions is too low for stellar kinematics (via absorption line measurements) to be used (Romanowsky et al. 2003; Coccato et al. 2009).

In addition to studying the kinematics (and thus matter distribution) of galaxies, PNe are also an integral part of the so-called cosmic distance ladder (Jacoby et al. 1992). Although PNe had already been proposed as possible "standard candles", it was not until the late 1970s that the so-called Planetary Nebula Luminosity Function (or PNLF, which describes the distribution of PNe observed in a galaxy as a function of luminosity; Jacoby 1980) began to be used as a distance indicator (e.g., Ford and Jenner 1978; Jacoby 1979; Jacoby and Lesser 1981). Ciardullo et al. (1989) demonstrated that the functional form of the PNLF showed a very sharp turnover and cut-off at $M^{\star}_{5007} \approx -4.5$ (where the absolute magnitude of a PN in the light of [O III] 5007Å is defined as $M_{5007} = -2.5 \log F_{5007} - 13.74$ and F_{5007} is the flux in the 5007Å line in erg s^{-1} cm^{-2} at a distance of 10 pc). While this behaviour is difficult to explain (see Ciardullo 2012, and Chap. 8), it is seemingly invariant across galaxies of all types and metallicities with only a small dispersion (Jacoby et al. 1992; Ciardullo and Jacoby 1992). Gesicki et al. (2018) recently claimed that the observed properties of the PNLF can be ascribed to stars with mass, 1.1 M$_{\odot}$ $\leq$M$\leq$2 M$_{\odot}$, which all reach similar luminosities in their models (employing new post-AGB evolutionary tracks of Miller Bertolami 2016, and assuming the nebula remains optically thick for a significant fraction of its evolution). However, observations of the PNe associated with the old stellar population present in the bulge of M31 strongly indicate that such stars are not responsible for the observed cut-off, with Davis et al. (2018) favouring a binary origin for the brightest observed PNe (a hypothesis which will be discussed in detail in Chap. 8).

1.3.3 Binary Central Stars and the Common Envelope Phase

As one may suspect from the title of this book, PNe and binary stellar evolution are inextricably linked—PNe trace the mass loss history of low- and intermediate-mass stars while binary evolution (particularly close-binary evolution) can have a spectacular impact on the mass-loss rate, energetics and morphology.

Perhaps the clearest and most dramatic way that binary evolution can affect the stellar mass-loss is through a common-envelope (CE) episode. The first mention of such a phase was presented by Paczynski (1976) though with reference to previous discussions with Ostriker (1973) and the PhD thesis of Webbink (1975). The CE phase and its impact on PN formation will be discussed in much greater detail in Chaps. 2 and 3 (as well as being mentioned in several other chapters). However, the CE phase as proposed by Paczynski (1976) is in essence an attempt to explain the formation of evolved close binaries, the orbital separation of which is smaller than the radius the more evolved star would have been expected to reach while a giant

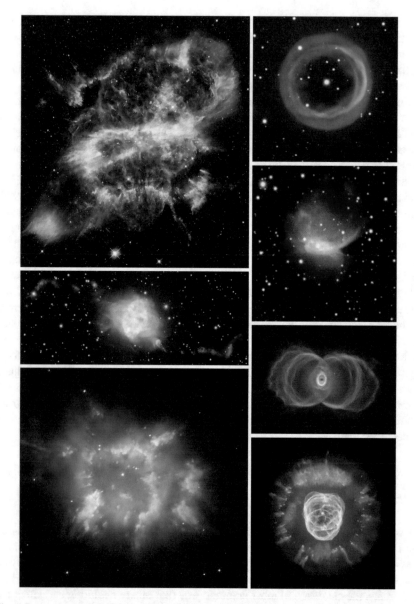

Fig. 1.7 A selection of PNe known to host binary central stars, highlighting the wide-range of observed morphologies (layout by Jakob Povel, Grantecan). Left column: NGC 5189, Fg 1, NGC 6326. Right column: Sp 1, Hen 2-428, MyCn 18, NGC 2392. Image credits: NASA, ESA and the Hubble Heritage Team (STScI/AURA); ESO/H. Boffin; ESA/Hubble and NASA; ESO; ESO; Raghvendra Sahai and John Trauger (JPL), the WFPC2 science team, and NASA; NASA, Andrew Fruchter and the ERO Team [Sylvia Baggett (STScI), Richard Hook (ST-ECF), Zoltan Levay (STScI)]

(i.e. at some point the giant would have engulfed its partner). In such systems, the more evolved star must (at some point during its giant evolution) have expanded enough to fill its Roche lobe and begin transferring material onto its companion. If the mass transfer is fast enough, the secondary will not be able to thermally adjust and will also fill its Roche lobe. As the primary continues to evolve, it will continue to overflow its Roche lobe leading to the formation of a CE of material surrounding both stars. With the two binary components now orbiting inside this CE, drag forces transfer orbital energy and angular momentum to the CE causing the two stars to spiral-in to shorter orbital periods—either unbinding and ejecting the CE to leave a short-period binary where one component is the core of the more evolved primary (now a pre-white dwarf [pre-WD]; Willems and Kolb 2004), or resulting in a stellar merger (similar to that of V1309 Scorpii; Tylenda et al. 2011). The first of these options is clearly a route towards the formation of a PN (with the ejected CE forming the nebular material, and the now exposed, hot pre-WD core being the source of the radiation required to ionise the nebula). Indeed, such a connection was noted by Paczynski (1976):

> Observational discovery of a short period binary being a nucleus of a planetary nebula would provide very important support for the evolutionary scenario presented.

The first discovery of such a short-period binary inside a PN came later that same year (Bond 1976, and many more have been discovered since—see Chap. 3). As such, given the relatively short life-times of PNe (a few tens of thousands of years), post-CE central stars are of particular interest in studying the CE as they are "fresh out of the oven", not yet having been able to adjust following the CE ejection (Jones and Boffin 2017a; Jones et al. 2019).

Transversely, post-CE central stars are also of particular interest in understanding the formation of PNe as CE evolution offers a clear route towards providing the axisymmetry required by the GISW theory to produce bipolar nebulae (see Sect. 1.1). As approximately 80% of all PNe show some deviation from sphericity (Parker et al. 2006), ranging from slightly extended ovoids through to bipolar hourglasses and irregular or multi-polar (see Figs. 1.1, 1.2 and 1.7 for some small demonstration of the observed variation in PN morphologies), understanding the processes which lead to the formation of such a wide-range of morphologies represents an important challenge in PN study (Balick and Frank 2002). Binarity is an attractive candidate to explain these variations due to the relatively high binary frequency on the main sequence (Raghavan et al. 2010) as well as the multiple ways in which such systems can evolve (depending on initial orbital and stellar parameters), leading to varying amounts of mass transfer (e.g., Miszalski et al. 2013a; Löbling et al. 2019), enhanced mass loss (Tout and Eggleton 1988; Sabach and Soker 2018), the launching of jets (Boffin et al. 2012b; Tocknell et al. 2014), as well as any number of other phenomena all of which can have a dramatic impact on the observed PN (see Chaps. 3 and 4 for further discussion in the context of short- and long-period binaries, respectively).

Chapter 2
The Common Envelope Phase

2.1 Mass Transfer in Binary Systems

Stars in a binary system may interact in various ways, depending on their evolutionary stage and on their separation. One of most interesting effects is when mass is transferred between the two objects, as this can have very dramatic consequences. Mass transfer can basically take place through two processes: when one of the components is losing mass via a stellar wind or when it becomes so large that some of its mass becomes gravitationally unbound. Stellar wind is important for two distinct kinds of stars: either hot, massive stars which have very strong radiatively-driven winds, or evolved low- and intermediate-mass stars. The first case explains systems such as massive X-ray binaries in which an O/B (or Wolf–Rayet) star transfers mass to a compact companion. The other case involves mostly stars that are on the red giant branch (RGB) or asymptotic giant branch (AGB), which lose a large fraction of their mass via stellar winds before ending their lives as white dwarfs (WD). This latter case explains systems such as ζ Aurigae systems and symbiotic stars, as well as playing a key role in the formation of stars polluted in carbon and s-process elements, that is, Ba or S stars (see Chap. 4). A detailed description of this mode of mass transfer in given in Boffin (2014; 2015, and refs. therein) and will not be consider further here.

Stars have stellar winds independently of being single or in binary systems—although the presence of a companion could reinforce the mass loss (see, e.g., Tout and Eggleton 1988). There is, however, another mode of mass transfer that is related to the particular shape of the equipotentials in a binary system, due to the combined effect of the gravitational forces of both components and their rotation around the centre of mass. In the ubiquitously used Roche model (Kopal 1959), where the two stars are approximated by synchronously-rotating points[1] in circular orbits, one can define so-called Roche lobes that represent the largest regions around a star where

---

[1] Or at least are assumed to be very centrally condensed.

© The Author(s), under exclusive license to Springer Nature Switzerland AG 2019
H. M. J. Boffin and D. Jones, *The Importance of Binaries in the Formation and Evolution of Planetary Nebulae*, SpringerBriefs in Astronomy,
https://doi.org/10.1007/978-3-030-25059-1_2

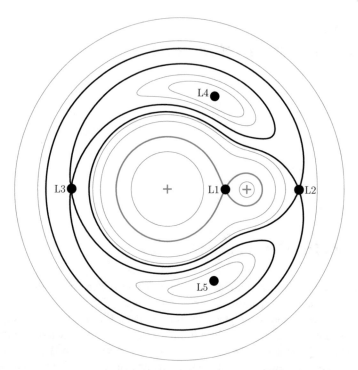

Fig. 2.1 A diagram showing lines of equipotential (including the Roche lobes, highlighted in red) as well as the five Lagrangian points for a system with a mass ratio of 5. The centres of mass of the two stars are marked with red +, with the more massive component being the leftmost

matter is still bound gravitationally to it (see Fig. 2.1). Such a model is of course extremely simplistic, as stars in binary systems will not necessarily be synchronous, they may fill a large portion of their Roche lobes, they may lose mass via stellar wind, and the orbit may be eccentric. Modern examples of trying to adapt the Roche model to various effects are found in, e.g., Dermine et al. (2009), Jackson et al. (2017), Hamers and Dosopoulou (2019), while it was shown half a century ago that asynchronism does not play an essential role in modifying the Roche lobe. However, in most cases, one needs to perform numerical simulations to represent more realistically all effects at play (e.g., Lajoie and Sills 2011; Deschamps et al. 2013).

The Roche model predicts the existence of several regions—the Lagrangian points—that play a specific role (e.g., Kopal 1959). In particular, the inner Lagrangian (or L_1) point, in between both stars, is important as it marks the place where matter from one of the stars will be able to move to the other one. This saddle point defines the Roche lobes, which are the equipotential surfaces around each star that touch each other at L_1. The Roche-lobe radius is a fraction of the orbital separation that depends on the mass ratio between the two components; it is larger for the most massive star in the system. In the approximation of Eggleton (1983)—estimated on an Apple-II-plus computer that was gifted to Eggleton by his wife!—the Roche lobe radius, r_L, is given by

$$r_L = a \frac{0.49 q^{2/3}}{0.6 q^{2/3} + \ln(1 + q^{1/3})} \equiv a f_L(q), \qquad (2.1)$$

where a is the separation between the two stars,[2] while q is the mass ratio.[3] The above relation is valid over the full range of q, and gives for a system with $q = 0.5, 1, 2, 5$, $r_L/a = 0.32, 0.38, 0.44, 0.52$, respectively.

The Roche lobe represents a critical equipotential that limits the possible size of a star. If a star becomes so large that it fills its Roche lobe, two things happen: first, the star will adopt the shape of the Roche lobe, i.e. will be pear-shaped. Secondly, it will start to transfer mass to its companion, as any matter that lies outside the Roche lobe will no longer be bound to the star, but will be captured by the companion via L_1. This is known as **Roche-lobe overflow** (RLOF). In addition, as the star fills a larger and larger fraction of its Roche lobe, tidal forces become more and more effective, and the system should thus become synchronised and circularised, thereby fulfilling two of the assumptions of the Roche model! If the star presents a strong stellar wind, whose velocity is smaller or comparable to the orbital velocity (as may be the case for AGB stars even in relatively wide orbits[4]), the wind will gradually fill the Roche lobe and will thus be transferred to the companion also via L_1, instead of via a canonical Bondi-Hoyle-Lyttleton wind accretion (e.g., Boffin and Anzer 1994). This is the so-called wind Roche lobe overflow (WRLOF; Theuns et al. 1996; Nagae et al. 2004; Jahanara et al. 2005).

The existence of the Roche lobes allows astronomers to define different classes of binary systems, depending on the relative sizes of the stars with respect to their Roche lobes. Thus, when both stars are well within their Roche lobe, the system is called *detached*. In this case, mass transfer will only take place via stellar wind. If one of the stars (generally the initially most massive as this is the one that evolves fastest) fills its Roche lobe, transferring mass to its companion via the inner Lagrange point, the system is called *semi-detached*. When both stars (over-)fill their Roche lobes, the envelopes of both stars are touching and the system is called in *contact*. A nice, albeit outdated, review of the importance of such classification is provided by Paczyński (1971), including the explanation of the Algol paradox.

A star can fill its Roche lobe during the various stages of its evolution, depending on the separation between the two components (hence, the orbital period). Roughly speaking, if the mass transfer occurs while the star is on the main sequence, astronomers talk of a Case A RLOF. If it is on the RGB, we talk of Case B, and if it is

---

[2]a is thus generally taken as the semi-major axis, but in case the orbit is eccentric, it may be necessary to use the periastron distance, $a(1 - e)$, where e is the orbital eccentricity—even if this is not self-consistent as the Roche model is formally only valid for circular orbits.

[3]Here, the mass ratio is defined as the ratio of the mass of the star, of which we want to measure the Roche lobe, to that of its companion.

[4]This would be even more true if the acceleration zone of the wind is located at several stellar radii (as is true for most AGB stars) and thus may be outside the Roche lobe itself.

while the star is on the AGB, it is a Case C (Lauterborn 1970). From Case A through to Case C, each takes place at longer and longer orbital periods—see Chap. 4 and Eggleton (2006).

2.2   Common Envelope Evolution in Brief

The mass transfer during RLOF can take place on a nuclear, thermal or hydrodynamic time scale, depending on the mass ratio and the response of the star filling its Roche lobe to the mass loss (Eggleton 2006; Ivanova 2015). A star with a substantial radiative envelope shrinks when undergoing mass loss, while a star with a deep convective envelope (such as a red giant) may expand. This could potentially lead to a runaway process (a positive feedback loop), as the more a star loses mass, the more it expands, and thus the more it overflows its Roche lobe, leading to yet further mass loss!

Benson (1970) was perhaps the first to realize that the companion to a mass-transferring star in a binary system would not appreciate the mass dumping. Driven out of thermal equilibrium, the companion expands and fills its own Roche lobe: a contact system will then arise (see also Yungelson 1973; Kippenhahn and Meyer-Hofmeister 1977; Neo et al. 1977). This led Paczynski (1976) to explore the outcome of such a phase, noting that due to the mass transfer instability, the red giant's envelope was not synchronously rotating with the orbital motion of the companion. He followed suggestions by Ostriker (priv. comm. to Paczynski in 1973) and Webbink (1975) to propose that such a non-corotating *common envelope* (CE), which encompasses both stars, may be Nature's way to produce cataclysmic variables with orbital periods of a few hours from binaries with very large initial orbital periods. The lack of corotation creates strong gravitational torques that force the companion to spiral-in down through the red giant's envelope, leading to a rapid decay of the orbit (Taam 1994). Paczynski (1976) in particular explained the existence of the binary system V471 Tau, containing a $0.8 \, M_\odot$ white dwarf and a $0.9 \, M_\odot$ K-type main-sequence star in a 12.5-hour orbit, by common envelope evolution, as in order to accommodate the asymptotic giant branch progenitor of the WD, the initial period had to be about 10 years. The idea was that when the AGB filled its Roche lobe, mass transfer occurred on a dynamical time scale (which is of the order of the initial orbital period) and a common envelope formed, due to the response of the secondary. Inside the CE, the main-sequence star spiraled-in towards the AGB's core, due to drag forces. This spiral-in was eventually halted, the end product being a close binary inside an expanding envelope, i.e. a planetary nebula with a close binary nucleus—a link that was also previously made by Vauclair (1972) to explain this same object (see also Livio and Soker 1988; Iben and Tutukov 1993). A cartoon schematic of the CE showing the onset of the Roche lobe overflow, as well as the formation and ejection of the envelope is shown in Fig. 2.2.

It is now thought that the CE phase is critical to explain many different kinds of systems, such as X-ray binaries, the progenitors of Type Ia supernovae and gamma-ray bursts, as well as binary black holes that have recently been detected as generating

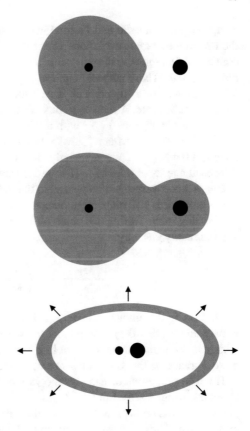

Fig. 2.2 A cartoon sketch
highlighting the key phases
of the CE process, from the
onset of Roche lobe overflow
(top) through the engulfment
of the companion (middle),
culminating in the ejection of
the CE and the shrinking of
the orbit (bottom)

gravitational waves. In some cases, the result of the CE could be a merger. For recent
reviews on the CE process and its implications, see, e.g., Webbink (2008), Ivanova
et al. (2013), Ivanova (2015, 2017), De Marco and Izzard (2017).

If there is no doubt that the CE occurs in Nature—there is no other way to explain
binary systems containing a white dwarf with orbital periods below a few days—
many things remain, however, unclear, and this explains why the CE is currently one
of the least understood processes in astrophysics, despite its importance. Among the
things to clarify are

- What is the critical mass ratio that would lead to the CE phase?
- How is the envelope ejected and what decides the final separation between the two
  components (or whether they merge)?

2.3 Dynamical Mass Transfer

Paczyński and Sienkiewicz (1972), following up on the work of Morton (1960)
which applied to radiative stars only, showed that the mass transfer takes place on
a "dynamical" time scale, when the mass-losing component is a red giant. They

computed the adiabatic and thermal evolution of the radius and compared it with the Roche lobe radius evolution. The important parameter is the ratio between the core and stellar masses, as well as the mass ratio between the two components of the binary system. These authors found in a simplified case, that the giant would lose 60% of its mass on a "dynamical" time scale and a further 13% of the initial mass on a thermal time scale. A more rigorous calculation for a binary comprising a Roche-lobe filling red giant of 1 $M_\odot$ with a 0.306 $M_\odot$ core, with a 0.67 $M_\odot$ companion, was then performed. After a "slow" phase of mass transfer of 2.3×10^4 yrs, the red giant lost 0.4 $M_\odot$ in a few years only, then went through a thermal mass transfer till it reached a mass of 0.36 $M_\odot$. The reason why *dynamical* was put in quotes is that, although this is very rapid compared to the thermal time scale of the star (5,000 yrs in this case), it is still quite large compared to the real dynamical time scale, which is a few days only. During this phase of fast mass transfer, the mass transfer rate was approximately given by

$$\frac{dM_1}{dt} = -23 \left(\frac{\Delta M}{M_\odot} \right)^{2.5} \frac{M_\odot}{yr},$$

where ΔM is the amount transferred. These calculations are of course simplified, in particular in that they assume that the mass transfer is conservative. If it is not, with mass and angular momentum lost from the system, the evolution can be even more dramatic, as the Roche lobe size will likely shrink even faster in this case (Tout 2012). The properties of cataclysmic variables indicate that such losses are to be expected (Ritter 1976). Lin (1977) studied the importance of the mass and angular momentum loss on the stability of the mass exchange, which depends mainly on the mass ratio and how the radius of the mass-losing star reacts to the mass loss. It was found that if the star reacts to mass loss by increasing in radius—as some red giants likely do—then the mass transfer is unstable if the mass ratio is greater than one.

A Criterion for the Stability of the Mass Transfer

More generally, the occurrence of a dynamical RLOF, which is the prerequisite for a CE to occur, relies on the comparison between the evolution of the Roche lobe radius and the evolution of the stellar radius with respect to mass loss. As described above, for red giants, the star may expand when losing mass. Thus, unless the Roche lobe's radius expands even faster than this, the process will be dynamically unstable. In the fully conservative case, the Roche lobe's evolution is only a function of the mass ratio, and it is thus the latter that will define whether a dynamical mass transfer occurs or not. The initial critical value for the mass ratio, q was thought to be of the order of 2/3 or 1. As the more massive component of a binary system evolves the fastest and thus reaches first the RGB and AGB first, it is generally thought that the mass ratio at the beginning of the mass transfer will be greater than one, and thus the CE is inescapable. However, the existence of systems containing a WD with orbital periods of 100–1,000 d (see Chap. 4) indicates that this cannot always be the case. Indeed, red giants lose much mass by stellar wind and it may well be that by the time a star fills its Roche lobe, the mass ratio has been reversed. This would be particularly plausible if binarity *enhances* mass loss (Tout and Eggleton 1988).

To determine if mass transfer will be unstable, one needs to compare how the radius of the star responds to the mass loss, and the way that the Roche lobe radius changes during the mass transfer. Thus, one generally uses the mass-radius exponents, linking the donor's radius, R_d, to the donor's mass, $R_d \propto M_1^{\zeta_*}$ and similarly for the Roche lobe radius dependence, $R_L \propto M_1^{\zeta_L}$. Then, **the mass transfer will be unstable if** $\zeta_* < \zeta_L$.

The value of ζ_* depends on the kind of mass transfer time scales one considers (Hjellming and Webbink 1987). A fully convective star can be approximated as a polytrope of index 1.5, which is characterized by $\zeta_* = -1/3$, leading to the general statement that such a star responds to mass loss by increasing in size, thereby leading to unstable mass transfer. While such a model is appropriate for a fully convective low-mass main-sequence star, red giant and asymptotic giant branch stars are more complicated and one needs to use either a condensed polytrope approximation or, better, full stellar models. In that case, we can see that ζ_* becomes larger than the value indicated above, and is often positive. A good approximation for red giants (Soberman et al. 1997) is:

$$3\zeta_* = 2\frac{q_c}{1 - q_c} - \frac{1 - q_c}{1 + 2q_c}$$

with $q_c = \frac{M_{1,c}}{M_1}$, the relative fraction of the core mass of the donor star. Thus, ζ_* becomes positive for $q_c > 0.215$, which would be the case for an AGB star initially less massive than $\approx 4\ M_\odot$.

Similarly, the value of ζ_L depends on whether the mass transfer is fully conservative or liberal (Eggleton 2000). In the fully conservative case, $\zeta_L \approx 2.13q - 1.67$, with $q = M_1/M_2$. This would allow to define that the mass transfer is unstable if $q > \frac{\zeta_*}{2.13} + 0.788$. For $\zeta_* = 0$, this means $q > 0.788$. As by definition, the donor was *initially* the most massive, this should always be the case, unless it has lost a non-negligible amount of mass before the RLOF starts! An AGB star of 1 $M_\odot$ may have $q_c \approx 0.57$, which leads to $\zeta_* \approx 0.8$, much different from the canonical $-1/3$ value. However, even in this case, mass transfer would be unstable for $q > 1.16$, which still encompasses most cases. For liberal mass transfer, the derivation is much more complicated (Soberman et al. 1997; Ivanova 2015), but it leads to smaller values of ζ_L, meaning that this makes it easier to have a stable mass transfer.[5]

The above derivation assumes hydrostatic equilibrium. However, as shown by Woods and Ivanova (2011), in real stars, the mass removal from the outer layer of the donor will perturb the hydrostatic equilibrium of the star, to which it responds on

---

[5]In addition to the fact that the mass ratio will increase as well if the donor loses mass before starting the RLOF.

its dynamical timescale. This has drastic consequences for the stability of the mass transfer. As an example, the authors considered a 5 $M_\odot$ giant donor and a core mass of $M_c = 0.856$ $M_\odot$. In the above approximation, one would predict an unstable mass transfer for any mass ratio $q > 0.75$. However, using the binary stellar-evolution code STARS/ev, the authors found that the mass transfer was stable for mass ratios as high as $q = 1.47$. Thus, one can only make strong conclusions for the response of any given donor on a case by case basis, using a detailed stellar evolutionary code. In the same vein, Passy et al. (2012b) found that the response of a giant to the mass loss was not adiabatic, and that the outer layer of the giant's envelope has a local thermal timescale comparable to the dynamical timescale, allowing it to readjust thermally. In these cases no increase of the stellar radius with respect to its initial value is found. They conclude that the conditions for unstable mass transfer may need to be re-evaluated. Pavlovskii and Ivanova (2015) showed that the recombination energy in the super-adiabatic layer plays an important role in the donor's response to mass-loss, in particular on its luminosity and effective temperature. They concluded that the critical initial mass ratio for which a binary would evolve stably through the conservative mass transfer varies from 1.5 to 2.2, which is about twice as large as previously believed. These revisions to the initial criterion for the stability of mass transfer may explain the existence of many post-mass transfer systems with orbital periods in the range 100–2,000 days (see Chap. 4), which are hard to explain with standard models. However, as we know of many post-CE systems, such a dynamical mass transfer does take place and it is necessary to understand how this happens. It is also clear from the known binary central stars of planetary nebulae that have orbital periods small enough to be the outcome of the CE process (see Chap. 3) that most have a rather low mass companion, i.e. a K or M dwarf. This means that the mass ratio prior to the CE event must have been rather high, $q > 6$. This fact, already noted by Eggleton (2006), may be the main reason why a CE took place.

2.4   Common Envelope Evolution Formalism

When the mass transfer is dynamically unstable and a common envelope forms in which the companion spirals in, the outcome may be either a merger or a short-period binary system. The latter will happen if the energy transferred to the envelope was large enough to unbind it. Thus, a parametric approach is generally used to equate the fraction, α, of the energy available in the orbit to the binding energy of the envelope (Tutukov and Yungelson 1979; Webbink 1984; Livio and Soker 1988; Livio 1989).

The difference in the orbital energy before and after the CE is given by

$$\Delta E_{\text{orb}} = \frac{G}{2} \left(\frac{M_{1,f} M_{2,f}}{a_f} - \frac{M_{1,i} M_{2,i}}{a_i} \right), \tag{2.2}$$

where the indices i and f signify initial and final, while index 1 refers to the donor (i.e. the progenitor of the white dwarf) and 2 to its companion. A fraction $\alpha < 1$ of this change in orbital energy is then assumed to equal to the binding energy of the envelope, E_b,

$$E_b \equiv G\frac{M_1 M_{1,\text{env}}}{\lambda R_1} = \alpha \Delta E_{\text{orb}}, \tag{2.3}$$

with $M_1 \equiv M_{1,i}$, while R_1 the radius of the primary (donor) star, having an envelope mass $M_{1,\text{env}}$, and λ is related to the evolutionary stage of the primary. The exact value of λ is difficult to estimate[6] and some authors (see below) propose a parametric formulation as a function of the core mass of the primary. One should note, however, that *single stars* seem to be able to lose their envelope at the end of the AGB phase, at which point the binding energy must be null, i.e. λ is infinite!

Noting that the donor must have filled its Roche lobe radius at the start of the mass transfer, and assuming that the companion is gaining little to no mass, $M_{2,f} = M_{2,i} = M_2$, one can thus compute the ratio of the final and initial separations,

$$\mathcal{F}(a) \equiv \frac{a_f}{a_i} = \frac{M_c}{M_1}\frac{M_2}{M_2 + 2(M_1 - M_c)/(\alpha\lambda f_L(q))}, \tag{2.4}$$

where $M_{1,f} \equiv M_c = M_1 - M_{1,\text{env}}$ is the core mass of the primary with mass M_1. As one can see in the above equation, α and λ only appear through their product $(\alpha\lambda)$. This is why only the value of this product is estimated empirically, rather than attempting to derive individual values for α or λ (Nelemans and Tout 2005). Webbink (2008) provides a formulation for λ that depends only on the mass of the donor and the mass of its envelope, or, alternatively, on the relative fraction of the core mass of the donor star, $q_c = M_c/M_1$. It thus appears that $\mathcal{F}(a)$ is only a function of q_c, q and α:

$$\mathcal{F}(a) = q_c \left[1 + \frac{2q(1 - q_c)}{\alpha\lambda f_L(q)}\right]^{-1}. \tag{2.5}$$

It can be seen that the right-hand term of the right side of this equation is always much larger than 1, and Eq. (2.5) can therefore be approximated by

$$\mathcal{F}(a) \simeq 0.5\frac{q_c}{1 - q_c}\lambda(q_c)\frac{\alpha f_L(q)}{q}, \tag{2.6}$$

so that one can see that this ratio is directly proportional to α. Using an initial-final mass relation (IFMR) for white dwarfs (Cummings et al. 2018), and ignoring any mass loss or accretion prior to the CE event, one can compute the ratio $\mathcal{F}(a)$ as a function of the mass ratio. This is shown in Fig. 2.3 for a value of $\alpha = 0.3$ and various companion masses.

The value of α is determined by the energy transport and non-spherical effects. When the energy can be transported rapidly (compared to the orbital decay timescale), it becomes more difficult for the orbital energy to be used for the ejection of the

---

[6]As is the definition of the core of a red giant (Tauris and Dewi 2001), which is crucial to be able to define λ!

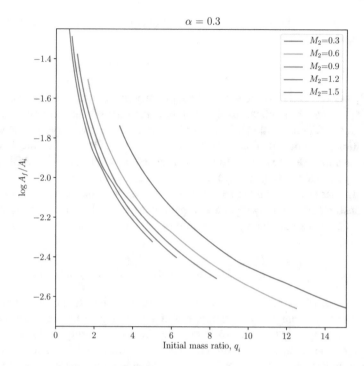

Fig. 2.3 The ratio of the final to initial separation as a function of the initial mass ratio, for a range of secondary masses, when assuming $\alpha = 0.3$ and Eq. 2.6

envelope (i.e. α is smaller). Numerical simulations also showed that mass ejection takes place preferentially in the orbital plane, which is the prerequisite to create a torus and the bipolar shape observed in planetary nebulae (Chap. 3). In that case, the material in the plane obtains velocities exceeding the escape velocity, thereby also reducing α. Taam and Bodenheimer (1989), for example, found in their two-dimensional hydrodynamical simulations of a 1 $M_\odot$ main-sequence star and a 5 $M_\odot$ red giant that more than about 3 $M_\odot$ were ejected in an equatorial outflow. They found that it was *"probable that the entire common envelope will be ejected and that the coalescence of the two cores can be avoided"* and that α was in the range between 0.3 and 0.6. In another simulation, Taam and Bodenheimer (1991) followed the late phase of the common envelope stage of a binary consisting of a 2 $M_\odot$ red giant and a 1 $M_\odot$ companion, which ended up with an orbital period of 1.2 days. They found that the crucial parameter defining the end state of the system was the density distribution in the AGB star, in the vicinity of the core—this was much more abrupt in the 2 $M_\odot$ AGB than in the 5 $M_\odot$ one, and this helps greatly in terminating the spiral in.

The α formalism doesn't consider any source or dissipation of energy other than gravitation and the binding energy of the envelope, as it is assumed that the CE evolution is very fast, more rapid than the thermal time scale of the envelope. However, there are also factors that could potentially increase α (Pastetter and Ritter 1989; Iben and Livio 1993):

1. The recombination energy of the ionisation zones;
2. Mass ejection by the AGB star;
3. Injection of fuel into the burning shells, leading to an enhanced nuclear energy production;
4. If the companion is a WD, nuclear burning on its surface;
5. Excitation of non-radial modes;
6. Pulsation and dust-driven winds.

Recently, the α-formalism has been extended (Nandez and Ivanova 2016; Ivanova and Nandez 2018) to incorporate the energy taken away from the envelope, as well as the recombination energy. The effect of the latter on the unbinding of the envelope has been and is still very much a matter of debate in the literature (Webbink 2008; Nandez et al. 2015; Ivanova 2018; Ivanova and Nandez 2018).

Assuming an initial period of about 10 yrs, a final post-CE period of half a day, typical masses of stars, gives $\mathscr{F}(a) \sim 2.6 \cdot 10^{-3}$. According to Fig. 2.3, this leads to a requirement of $q > 9$ for $\alpha = 0.3$. Equation (2.6) shows that the final separation increases with increasing q_c, increasing α and decreasing q (i.e., increasing companion mass). Thus, if there was mass and angular momentum loss prior to the mass transfer, this will lead to a larger value of q_c, a larger value of q, and to get the same a_f, a smaller α is needed.

2.5 Hydrodynamical Simulations

As we have seen, the formalism to deal with common envelope evolution is very parametric and far from being fully understood. Moreover, the mass transfer process is intrinsically three-dimensional. It is thus clear that a better understanding of this phenomenon will only come from multi-dimensional hydrodynamical (and possibly magneto-hydrodynamical) simulations, that incorporate very sophisticated physics. This is now becoming more and more attainable, although still very demanding in terms of computing resources such that simplifications have to be made. For example, many simulations assumed that the spiral-in was already ongoing (i.e., didn't try to follow how the envelope was created) or considered a red giant donor instead of an AGB star that would have a less bound envelope.

The first three-dimensional hydrodynamical simulations (Taam and Sandquist 1998) indicated that when the companion reaches the original surface of the mass-losing giant, the spiral-in phase accelerates and the orbital separation decreases abruptly on timescales of 50–100 days. After about 1 year, i.e. on the same order as the initial orbital period, the spiral-in is much slower, happening on timescales of 10 or so years. The final separation was found to depend on the evolutionary state of the red giant and on the mass ratio, with the final separation increasing for more massive red giant cores and more massive companions.

Using different 3D hydrodynamical codes, Passy et al. (2012a) looked at the rapid in-fall phase of the common envelope (CE) interaction of a sub-solar mass red giant and a companion star. The simulations showed that most of the envelope of the donor remains bound at the end of the simulations and the final orbital separation between the donor's remnant and the companion was larger than those observed in post-CE binaries. Among others, they therefore suggested that an additional source of energy was needed to eject the envelope. Following-up on this work with a more sophisticated hydrodynamical code, Iaconi et al. (2018) concluded that even if smaller final separations were obtained, the envelope could still not be fully ejected. They also showed that with a more bound envelope, all the companions in-spiral faster and deeper, though relatively less gas is unbound.

Ricker and Taam (2012) followed for five orbits the evolution of the common-envelope phase of a low-mass binary composed of a red giant and a lower mass companion. They found that during a rapid in-spiral phase, in which the orbital separation decreased by a factor seven, the interaction of the companion with the red giant's extended atmosphere causes about 25% of the CE to be ejected from the system, with mass continuing to be lost at the end of the simulation at a rate of ~ 2 $M_\odot \text{yr}^{-1}$. They also found that 90% of the outflow is contained within 30 degrees of the orbital plane. Later, Ohlmann et al. (2016) followed the interaction of a 1 $M_\odot$ compact star with a 2 $M_\odot$ red giant with a 0.4 $M_\odot$ core. They discovered that following the initial transfer of energy and angular momentum from the binary core to the envelope by spiral shocks, turbulent convection started to develop in the CE, thus altering the transport of energy on longer timescales. Nevertheless at the end of their simulation, only 8% of the envelope mass was ejected.

Chamandy et al. (2018) looked at the possibility of accretion onto the in-spiraling companion as an additional source of energy to unbind the envelope. They demonstrated that if a pressure release valve is available—for example in the form of jets—super-Eddington accretion may be common and this could also help unbind and shape the ejected envelope. This was followed by the study of Chamandy et al. (2019) who found that virtually all of the envelope unbinding in the simulation occurs before the end of the rapid plunge-in phase. During the subsequent slower spiral-in phase, energy continues to transfer to the envelope from the red giant and secondary cores. More recently, Reichardt et al. (2019) found that the mass transfer rates and gravitational drag in their simulation of a common envelope event are reasonably well approximated by the analytic values. Moreover, they saw that the L_2 and L_3 outflows observed during the Roche lobe overflow remain bound, forming a circumbinary disc that may or may not be disrupted by the common envelope ejection. They conclude that the observed wide range of post-common envelope planetary nebula shapes can be explained by varying degree of stability of the mass transfer.

Many of the previous simulations were looking at the rapid in-fall phase and had to be stopped before the envelope was completely unbound. This makes a comparison with observations very difficult. Recently, Frank et al. (2018) looked at the subsequent phase, by simulating the interaction of the fast wind and common envelope ejecta and found that this leads to highly collimated (i.e. jet-like) bipolar outflows. Similarly, García-Segura et al. (2018) demonstrated that apparently multi-lobed structures (such as those found in Abell 65; Huckvale et al. 2013) can be produced without the need for multiple ejection events, simply as a result of the flow of photo-evaporated gas from the equatorial regions. Collectively, simulations demonstrate that common envelope evolution has a major impact on the shaping of planetary nebulae.

2.6 Double-Degenerate Systems

In addition to the traditional post-CE systems that contain a white dwarf and a non-degenerate companion, there also exist a sub-class of close binaries that contain two degenerate stars, the double-degenerate systems. As we will see in Chap. 3, there is in fact an apparent over-abundance of such systems compared to population synthesis models (Boffin et al. 2012b; Jones and Boffin 2017a). Such systems must have gone through two successive mass-transfer episodes. In the first one, the system most likely avoided the CE or at the very least the effect of the CE must have been very small, because the remaining system must have been still wide enough to leave space for the secondary to expand to the RGB or AGB phase. In the second mass transfer event this was not the case. As mentioned earlier, the existence of symbiotic and other peculiar red giant systems composed of a red giant and a white dwarf with orbital periods of several hundreds or thousands of days (Chap. 4) shows that such stable episodes of mass transfer are not uncommon, even though the details are still far from being understood. Their explanation may lie in the fact that the initial mass ratio was perhaps not very high as discussed previously (see also Jones 2019). The existence of such systems has important consequences as these double-degenerate systems may be gravitational wave emitters that could be detected by LISA (Nelemans 2018) or could be the progenitors of Type Ia supernovae (Santander-García et al. 2015).

2.7 Grazing Envelope Evolution

Although the common envelope process has been the most studied in the literature to explain the formation of post-mass transfer short-period close binaries, we cannot finish this chapter without mentioning an alternate scenario, proposed by Soker (2015)—the grazing envelope evolution (GEE). This mechanism is based on the fact that jets launched by the companion can facilitate envelope ejection. Thus, as described in the original paper,

> In the GEE the system evolves in a constant state of just entering a CE phase. Namely, parts of the giant envelope overflow the Roche lobe in a large volume around the first Lagrangian point L1, but jets (or disk wind) launched by the secondary star prevent the formation of a CE

As many post-mass transfer central stars of PNe show the presence of jets (Chaps. 3 and 5), there may be good reason to think that GEE took place in such systems. One particular case where the GEE could have taken place, as suggested by Soker (2015), is in one of the longest of all post-CE central binary systems of PNe, NGC 2346 ($P_{orb} \sim 16$ days; Brown et al. 2019).

A major difference between the GEE and the CEE is in its duration—while the duration of the final CE phase is thought to last between days and some months at most, the overall GEE may last tens or even hundreds of years.

Chapter 3
Close-Binary Stars in Planetary Nebulae

3.1 Searches for Close-Binary Stars

3.1.1 The Beginnings

Following the discovery of the first close-binary central star by Bond (1976), through careful comparison of the Perek–Kohoutek Catalogue of Galactic Planetary Nebulae (Perek and Kohoutek 1967) and the General Catalogue of Variable Stars (Kukarkin et al. 1971), the search for other systems was principally carried out by Howard E. Bond, Albert D. Grauer and Robin Ciardullo—leading some to call it the Bond–Grauer–Ciardullo survey (De Marco et al. 2008). However, this was not a survey in the modern sense of a uniform series of observations carried out with the same instrument and same telescope, but rather a concerted effort over three decades to photometrically follow the central stars of some ~100 PNe (often selected due to previous indications of variability or at least biased towards low surface brightness nebulae for ease of photometry; Grauer and Bond 1983), using whatever telescopes and instruments were available (principally 1m-class telescopes equipped with one- or two-channel photomultiplier-based photometers).

In spite of the limitations of the Bond–Grauer–Ciardullo survey, it provided the first wave of detections of post-CE central stars totalling roughly ten by the year 2000 (Bond 2000)—though in one case the photometrically variable star has since been shown to be a chance superposition rather than the true central star (SuWt 2; Jones and Boffin 2017b). Simultaneously, the same authors and others were taking an interest in PNe with central stars which appeared abnormally bright or red, where the spectral type indicates a star that would not be capable of ionizing the surrounding nebula and is therefore likely to form a binary pair with the nebular progenitor. These systems are often referred to in the literature as *peculiar* central stars (Aller et al. 2015). In several cases, studies with the International Ultraviolet Explorer space telescope revealed the presence of the hot primary component of the binary, which is dominated in the optical by the cooler companion (e.g., Tyndall et al. 2013). Furthermore, photometric

© The Author(s), under exclusive license to Springer Nature Switzerland AG 2019   27
H. M. J. Boffin and D. Jones, *The Importance of Binaries in the Formation and Evolution of Planetary Nebulae*, SpringerBriefs in Astronomy,
https://doi.org/10.1007/978-3-030-25059-1_3

studies revealed that most also displayed photometric variability on timescales of a few days, similar to those of the bona-fide post-CE systems. However, in these cases the variability was in fact due to rotation of the cool companion rather than the typical irradiation or tidal distortions (which are the prime causes of photometric variability in the post-CE systems, see Sect. 3.1.3). Only one of the so-called *peculiar* central stars has been found to be a post-CE system, rather than a wider binary which has avoided a CE phase (see Chap. 4 for a more detailed discussion of the other *peculiar* central stars), that of NGC 2346. The central star of NGC 2346 is a somewhat unique case in that it was the first to have its period measured spectroscopically rather than photometrically (Méndez and Niemela 1981), is one of the longest period post-CE systems known and also presents with the most massive companion known (Brown et al. 2019). We reserve further discussion of its properties (and those of the rest of the post-CE central star population) for Sect. 3.2.

3.1.2 Modern Photometric Surveys

The first detection of a binary central star by a modern, wide-field photometric survey came from the Massive Compact Halo Object (MACHO) survey, designed to detect micro-lensing events but equally adept at recovering photometric variability of other kinds (Alcock et al. 1999). Lutz et al. (1998, 2010) reported that while for almost a quarter of the PNe in the MACHO fields, the data were insufficient to detect variability, the observations of Hf 2-2 did reveal a periodic variable with a period of 0.39 days consistent with a post-CE binary (since confirmed by Hillwig et al. 2016a).

A much greater leap forward came from the Optical Gravitational Lensing Experiment (OGLE; Udalski et al. 2008) which benefitted from observing the Galactic bulge in the I-band (MACHO observed in its own non-standard filters: blue and red). The I-band has proven particularly optimal for studying the variability of central stars for two main reasons:

1. Of all the standard optical, broad-band filters (e.g., Bessell 1990), the I-band covers the fewest bright, nebular emission lines, thus generally minimizing contamination in the photometry of the central star (Jones et al. 2014). As discussed in Miszalski et al. (2009a), the non-standard I-band of OGLE can be contaminated in PNe which display particularly strong nebular continuum or where the [SIII] lines at 9069Å and 9532Å (the second of which lies outside many I-band filters which generally cut-off at $\sim$9000Å) are particularly bright. However, in these cases bluer filters would almost certainly display even greater levels of contamination from the lines of the Balmer series or [O III] (see Fig. 3.1 and Sect. 3.1.3).
2. The amplitude of the irradiation effect is generally larger in the I-band than in other optical bands (see Sect. 3.1.3; and also De Marco et al. 2008), thus making such effects more easily detectable.

This means that through careful cross-matching of OGLE sources with the positions of PNe from the literature (accounting for non-PN contaminants in the literature sample), Miszalski et al. (2009a) were able to identify 21 new periodically variable central stars consistent with being binaries, more than doubling the known sample.

Following the discoveries from the OGLE survey, it was possible to begin to make statistical links between different nebular morphologies and the presence of a detectable, close-binary nucleus. Miszalski et al. (2009b) combined imagery from the literature with newly acquired narrow-band, emission-line images of some 33 PNe with close-binary central stars in order to probe the morphological trends amongst the sample. They showed that a significant fraction of the sample displayed canonical bipolar morphologies (perhaps as high as 60% once inclination effects were accounted for), just as would be predicted from models of CE-driven PN formation (see Chap. 2 and Nordhaus and Blackman 2006). Similarly, a higher than average fraction of the nebulae also displayed low-ionisation structures such as filaments or knots, or bipolar jets. These morphological features have since been used by the authors and collaborators as highly successful selection criteria for targeted searches for binarity (see, e.g., Miszalski et al. 2011a, c; Boffin et al. 2012b; Jones et al. 2014). By pre-selecting targets based on their morphologies (rather than, for example, the ease of photometry) and the novel-use of narrow-band continuum filters (e.g., Hβ continuum filters—the bandpass of which contains no bright nebular emission lines; Jones et al. 2015, 2019) on larger aperture telescopes (larger apertures are obviously required to observe faint stars through narrow-band filters while still maintaining reasonable temporal resolution), the authors were able to detect $\sim$10 new close-binary central stars.

The contribution of other modern surveys (beyond OGLE) has been rather limited, with most wide-field variability surveys being poorly suited to the task of searching for binary central stars, many being designed to hunt for exoplanets around bright stars. Such surveys generally present with large plate-scales meaning observations of PNe often suffer from contamination either through crowding (many PNe reside in the Galactic bulge where the stellar density is highest) or the use of broadband filters where nebular contamination is most problematic. In spite of this, the Super Wide-Angle Search for Planets (SuperWASP; Pollacco et al. 2006) was used to identify the 6.4 day variability period of the central star of LoTr 1, although this is, in fact, the rotation period of the giant secondary in a long-period binary system (see Chap. 4). The Kepler space mission (Borucki et al. 2010), however, has been relatively successful in detecting binary central stars—although it too suffers from the same drawbacks mentioned above, but this is somewhat counter-balanced by its much greater sensitivity with respect to ground-based counterparts. In the first Kepler campaign, De Marco et al. (2015) discovered that four out of five PNe with usable data presented photometric variability consistent with a binary evolution. Although only three of these are consistent with a surviving binary (they conclude that the central star of NGC 6826 is most likely the product of a merger), it is important to note that only one of the four would have been easily detected as variable from the ground (given that the variability amplitudes of the others were only on the order of a few milli-mag). Indeed, the central star of AMU 1 is, to-date, the only central star found to vary photometrically due to relativistic (Doppler-)beaming effects (Zucker et al. 2007), an effect so small it is exceptionally difficult to detect from the ground (see Sect. 3.1.3.4 for further discussion).

3.1.3   The Limitations of Photometric Surveys

Given that the vast majority of known binary central stars were discovered through photometric variability, it is important to understand the limitations such photometric monitoring presents and what sort of systematic biases this could impose upon the known sample. As already mentioned in Sect. 3.1.2, contamination of the photometry by bright and irregular nebular background can be extremely limiting. In the best case scenario, this contamination has the effect of a "third light" in the system, acting to reduce the observed amplitude of any variability thus making it harder to detect. In the worst case, it may even introduce spurious variability, depending on observing conditions (seeing, sky background) and the methodology employed to photometer the central star. The level of nebular contamination depends strongly on the filter employed, with different filters encompassing different nebular emission lines within their bandpass (see Figs. 3.1 and 3.2). For example, the R filter of the Bessell system (Bessell 1990) contains the brightest line of the Hydrogen Balmer series (Hα) as well as bright lines of [N II] and [S II] (see Fig. 3.1). The B filter covers Hβ as well as [O III] 5007Å in the wings of its bandpass. [O III] 5007Å also appears more prominently in the V filter. This means that B, V and R filters are generally relatively poor choices when searching for central star variability.

To further understand the limitations of photometric surveys for binarity, we must consider the possible sources of photometric variability and what relation they have to the parameters of the binaries themselves. It is important to highlight that the binary systems are not resolved and, as such, by photometric variability we are considering the variations in the observed (i.e. from the perspective of the observer) brightness of the system as a whole. In general, the binary central stars of PNe (bCSPNe) can be observed to vary photometrically as a result of four different effects. Here, we will discuss each effect separately although it is quite possible that a given system may exhibit more than one or even all in its light curve.

3.1.3.1   Eclipses

Perhaps the most obvious source of photometric variability is the eclipse of one star by the other resulting in a periodic drop in the observed brightness. For an eclipse to be observed the configuration of the system must be such that one star passes in front of the other. This is clearly dependent on the inclination of the binary orbital plane (where the convention is to define an orbital plane directly in the line of sight as having inclination equal to 90°), however it is also a function of the stellar radii

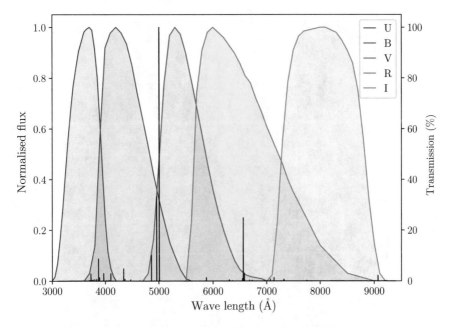

Fig. 3.1 A VLT-UVES spectrum of the PN IC 4776 (originally presented in Sowicka et al. 2017) with the transmission curves of standard Bessell filters overlaid (Bessell 1990)

and orbital separation—larger stars in tighter orbits will display eclipses at lower inclinations, while small stars in wider orbits will only eclipse at inclinations very close to 90°.

The configuration of the system can also lead to difficulties in eclipse detection— clearly, systems which spend less time eclipsed (longer orbital periods, smaller stars, lower inclinations) are less likely to be recovered by non-continuous monitoring. Similarly, if a system is only observed during a short observing run of, say, just a few nights (rather than an extended survey like OGLE) then there is a clear reduction in the likelihood of observing an eclipse in a longer period system. Furthermore, depending on the parameters of the binary and the photometric band in which it is observed, it is possible that only one eclipse may be detected.[1] For example, if the companion to the hot pre-WD central star is a much smaller, colder star, the eclipse of the companion may be difficult to detect in bluer-bands where its contribution to the total flux of the system is minimal. Of course, the companion is still eclipsed, it is simply that higher signal-to-noise measurements will be required to observe the corresponding drop in brightness. As such, there may be cases where random sampling of the light curve does not lead to the detection of a binary just because none of the observations coincided with a primary eclipse.

---

[1]Eccentric systems may also present with only a single eclipse due to the orbital configuration. However, given that we are discussing post-CE binaries and that the CE phase acts to circularise the binary orbit, here we will choose to ignore this possibility.

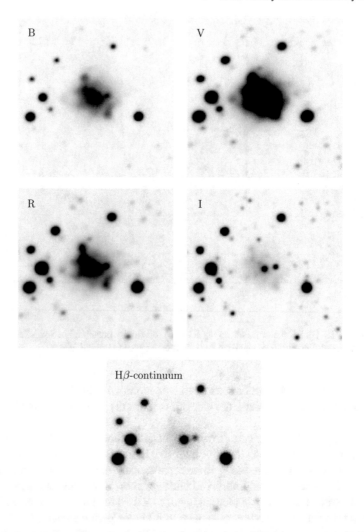

Fig. 3.2 NTT-EFOSC2 imagery of Hen 2-155 in the light of Johnson B, V, R and I, as well as
Hβ-continuum, highlighting the differing levels of nebular contamination visible in each band (each
image measures roughly 40"×40"). Ultimately, the Hβ-continuum filter was used by Jones et al.
(2015) to study the central binary

3.1.3.2   Ellipsoidal Modulation

Stars in close-binary systems are not spherical, rather they are distorted by the mutual
gravity of the entire system. In order to understand this effect, one must consider the
gravitational potential of a binary system. Close to the centres of mass of each star,
the surfaces of gravitational equipotential are approximately spherical while further
away the surfaces become elongated along the axis connecting the two centres of

Fig. 3.3 An example of a detached binary

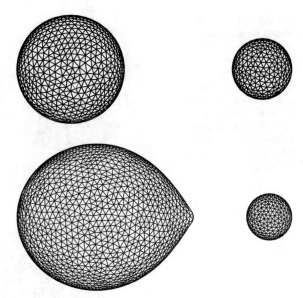

Fig. 3.4 An example of a semi-detached binary, where one component is filling its Roche lobe. In this case, the Roche lobe filling star is twice as massive as its companion

mass. The precise shapes of these surfaces of equipotential are a function of the stellar mass ratio and orbital separation. An example plot showing the configuration of various lines of gravitational equipotential for a hypothetical binary system is presented in Fig. 2.1.

It is now useful to define some important terminology used to describe binary stars.[2] Firstly, the smallest continuous surface of equipotential which encloses both stars is known as the Roche potential (marked in red in Fig. 2.1). It is often convenient to think of the potential as two lobes joined via the inner Lagrangian point (L1), where each lobe represents the volume within which material is bound to the star within that lobe. The Lagrangian points (also marked on Fig. 2.1) are the saddle points of the gravitational potential of the system with the first three Lagrangian points being local maxima, while the fourth and fifth Lagrangian points are local minima. While all five points are important in terms of orbital mechanics, those of primary interest in the context of close binary stars are the first, second and third. The first Lagrangian point is the point through which material is transferred between the stars (at the on-set of the CE, for example), while the second and third are the primary locations through which material is lost during the CE.

Binary systems where both stars are smaller than their Roche lobes are often referred to as 'detached' (see Fig. 3.3), while when one star is filling its Roche lobe the system is referred to as 'semi-detached' (see Fig. 3.4). When both stars are filling their respective Roche lobes, the system is classified as 'contact' (see Fig. 3.5).

Systems where one or both stars show strong deviations from sphericity, namely semi-detached and contact binaries, can present photometric variability throughout

---

[2]For completeness, we repeat here some of the notions defined in Chap. 2.

Fig. 3.5 An example of a contact binary, where both components are filling their respective Roche lobes, at an orbital inclination of 45°. The upper panel shows the system at quadrature (the phase of maximum projected surface area towards the observer) and at conjunction (the minimum of projected surface area)

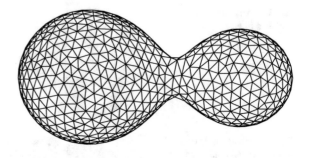

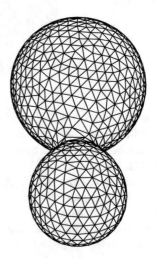

their orbits due to the differing projections of the stars. As the elongation of the stars is always along the axis connecting the stellar centres of mass, when the line of sight is parallel to that axis the stars present with a smaller projected surface area than when the axis is perpendicular to the line of sight. This effect is demonstrated in the two panels of Fig. 3.5, which shows a contact binary at quadrature and at conjunction. At quadrature, the full tear-drop shaped projection of both stars are clearly visible, while at conjunction the profiles of both stars are closer to spherical—even in this intermediate inclination case where essentially only the 'neck' of the binary (the point where the two stars join through the first Lagrangian point) is eclipsed by the star closest to the observer. Such systems are often referred to as ellipsoidal variables and the variability as ellipsoidal modulation (due to the roughly ellipsoidal shapes of the stars). Ellipsoidal modulation characteristically produces a roughly sinusoidal light curve with two minima per orbital period (one at each conjunction), which can be unequal in depth if the two stars are not identical. Furthermore, depending on the

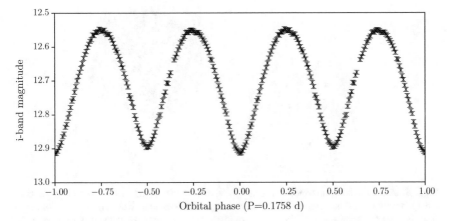

Fig. 3.6 An *i*-band light curve of the central star of Hen 2-428 folded on the orbital period of 0.1758 days (data originally presented in Santander-García et al. 2015). Note the unequal depths of the minima, which are a result of the slight difference in temperature between the two stars

orbital inclination, eclipses can also be superposed on this ellipsoidal modulation producing more pronounced deviations from a sinusoidal shape around the minima (Fig. 3.6).

Given that at least one star must be (close to) filling its Roche lobe for an ellipsoidal modulation to occur, the effect is only observed in the shortest period post-CE stars (as the Roche radius is a function of orbital separation and, therefore, period). Therefore, the technique is insensitive to periods longer than a few days unless one component is a giant (Nie et al. 2012; Prša et al. 2016). Furthermore, it is important to note that the amplitude of ellipsoidal modulation is generally rather small (≤ 0.05 magnitudes; Soszynski et al. 2004; Nie et al. 2012), as the star (or stars) must be very close to Roche-lobe filling in order for there to be an observable distortion in the stellar morphology. However, several binary central stars have been shown to present larger amplitudes (up to ~ 0.5 magnitudes) where both stars must be Roche-lobe filling (Miszalski et al. 2009a; Santander-García et al. 2015). These systems are generally thought to be double-degenerates, comprising two white dwarfs with relatively similar luminosities and temperatures (Hillwig et al. 2010). This idea will be discussed further in Sect. 3.2.

3.1.3.3 Irradiation

A particularly prevalent source of photometric variability in close-binary central stars is the so-called irradiation or reflection effect. In close binaries, the inward facing hemispheres of both stellar components are irradiated by flux from the opposing star. This incident flux can then either be scattered or heat the stellar surface (perhaps also being redistributed, see Wilson 1990; Budaj 2011; Horvat et al. 2019), with the collective result being that the irradiated (or day-side) hemispheres are more luminous

Fig. 3.7 A toy model of an irradiated binary, where the colour of the surface mesh represents the emergent flux from that element (with yellow being more luminous than red). Note the luminous 'hot spot' on the hemisphere of the cool star closest to the hot star—it is the differing projection (throughout the binary orbit) of this heated hemisphere which results in the sinusoidal variability characteristic of irradiated binaries

than the non-irradiated (night-side) hemispheres (see Fig. 3.7). The differing projection of the irradiated hemisphere throughout the binary orbit then leads to a periodic variation in the observed brightness of the system. Unlike ellipsoidal modulation, irradiation results in only a single minimum per orbital period (when the night-side of the irradiated star is in projection towards the observer). As such, when an irradiated system is also eclipsing, the eclipses will coincide with the minimum and maximum of the irradiated variation (again, unlike ellipsoidal modulation where minima and eclipses are always coincidental). The overall shape of the modulation, however, is roughly sinusoidal (with some strong dependencies on orbital inclination and limb darkening) in a similar fashion to that produced by ellipsoidal variability, meaning that the two can be difficult to differentiate in non-eclipsing systems without the aid of radial velocity observations to constrain the orbital period. An example light curve of an irradiated binary with intermediate inclination (i.e. non-eclipsing) is shown in Fig. 3.8.

The amplitude of an observed irradiation effect depends on a combination of factors including the orbital inclination and separation, stellar temperatures and radii, even on the nature of the stellar envelope (with radiative stars having greater albedos than those with convective envelopes; Ruciński 1969; Claret 2001). Simplistically, higher inclination binaries will present with greater irradiation effect amplitudes as the day-night divide of the companion lies perpendicular to the orbital plane. In closer orbits, the companion will receive more irradiating flux on its day-side surface leading to a larger irradiation effect. Similarly, larger companions intercept more irradiating flux, while hotter primaries emit more radiating flux—both leading to a greater observed amplitude. Additionally, the effect is colour-dependent, with redder bandpasses generally presenting with larger amplitude irradiation effects. The complex interplay between all these variables means that binary central stars with a wide-range of parameters, including both relatively long periods and low inclinations

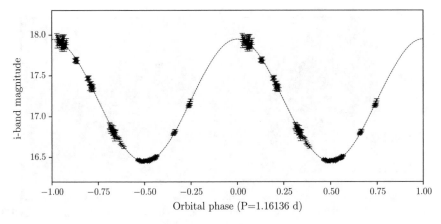

Fig. 3.8 An i-band light curve of the central star of the Necklace folded on the system's orbital period 1.16136 days (data originally presented in Corradi et al. 2011b). A sinusoidal fit is overlaid, highlighting the characteristic shape of the irradiation effect

(e.g. Sp 1 with a period of 2.9 days and an inclination of only 7–11°; Hillwig et al. 2016b), have been identified due to irradiation effect variability. However, there are clear limitations with extremely low inclination systems or those with periods greater than roughly 10 days presenting with no, or at least no detectable, variability due to irradiation (Jones and Boffin 2017a). Similarly, systems with very low mass companions will also be beyond reasonable detection limits as their observed brightness is dominated in all bands by the hot primary.

It is also important to note that irradiation can result in the production of bright emission lines in the atmosphere of the companion. This can be particularly useful when it comes to constraining the parameters of the system as these lines can be used to derive the radial velocity of the companion in spite of its faintness compared to the bright primary (see Sect. 3.2 for further discussion).

3.1.3.4 Relativistic Effects

As specific intensity is not a relativistic invariant, orbital motion causes a relativistic beaming of the light emitted from the components of a binary some times known as Doppler Boosting. This effect is also related to other effects including, for example, the Doppler shift of the stars' emergent spectra which may change the flux observed within a given photometric passband. This is an extremely low-level effect (amplitudes of $\leq 0.1\%$; Zucker et al. 2007) which is particularly difficult to detect from the ground (Shporer et al. 2010). However, its amplitude does not depend as strongly on the orbital period as irradiation or ellipsoidal modulation, meaning it is the dominant source of photometric variability in binaries with periods greater than approximately 10 days (Zucker et al. 2007). To date, only one binary central star has been detected

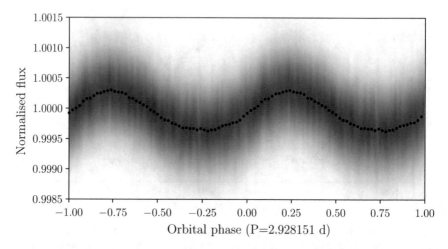

Fig. 3.9 The Kepler light curve of the central star of AMU 1 folded on the orbital period of 2.928151 days (data originally presented in De Marco et al. 2015). The binned light curve is shown as the circular points overlaid on a raster plot of all data points. As is characteristic of Doppler boosting, the light curve shows a single peak per orbital period around quadrature, unlike irradiation effects which show a single peak around conjunction and ellipsoidally modulated binaries which show two peaks per orbital period around quadrature

due to relativistic effects, that of AMU 1 shown by De Marco et al. (2015) to have an orbital period of 2.928 days using data from the Kepler space mission (see Fig. 3.9).

3.1.4 Spectroscopic Searches

While time-resolved photometry has proved the most efficient and successful way to detect close-binary central stars, several binary central stars have also been detected by spectroscopic methods. The spectroscopic equivalent to the time-resolved photometry described above would comprise time-resolved measurements of the radial velocities of one or both binary components. The obvious benefit of such an approach is that the technique can be sensitive to much longer period systems which will not present with any photometric variability (e.g. Miszalski et al. 2018a, b). However, it does present with several complications, namely:

- Radial-velocity monitoring necessitates a much larger investment, in terms of observing time per unit collecting area, than photometric monitoring. Simply put, obtaining a similar signal-to-noise in a sufficiently high-resolution spectrum (so as to be able to measure the changes in radial velocity) will always require a longer exposure time than for an equivalent photometric measurement (even using narrow-band filters). This is an obvious consequence of the spectral dispersion, as well as the fact that spectroscopy as a technique generally implies more components in the optical path and therefore a reduced transmission efficiency. Beyond the clear limitations in terms of total observing time that can be dedicated to this kind of spectroscopic monitoring (hours of observing time applied for generally out-number the available time several-fold at most modern observing facilities), there is also a limitation imposed by the orbital periods one hopes to probe. In order to adequately sample a binary radial velocity curve, the exposure time of a given observation must be significantly shorter than the orbital period otherwise the measurement will be "smeared". This implies that in order to probe the shortest periodicities and/or faintest central stars, it becomes necessary to use large telescopes where, again, observing time is generally very over-subscribed.

- While photometric monitoring efforts can also be adversely affected by nebular contamination (as described above), spectroscopic measurements are generally even more difficult. In most cases, the strongest absorption lines present in the spectrum of a (pre-)WD central star will be Hydrogen lines of the Balmer series which, unfortunately, are also some of the strongest nebular emission lines. While careful subtraction of the nebular emission (via, for example, interpolation of the nebular emission present either side of the stellar signature on a longslit spectrum) can prove effective, this is often made extremely difficult by the filamentary or knotty nature of the nebular emission.

- Radial velocity measurements can be made difficult by the presence of strong and/or variable stellar winds (which are often observed in these young pre-WD central stars). These can introduce additional signs of variability in the observed radial velocity curve beyond the orbital motion of a binary, thus even more observations are required in order to disentangle the periodic variability due to the binary orbit from quasi-periodic oscillations in the stellar wind profile (De Marco et al. 2004). Even in cases where wind variability may not be an issue, if there are no prior indications of the orbital period, it is often necessary to obtain twenty or more points (with random sampling) in order to unambiguously determine the true orbital period (Manick et al. 2015; Sowicka et al. 2017; Miszalski et al. 2019b)— with additional sources of variability this number can rise dramatically.

In spite of these difficulties, several binary central stars have been discovered via radial velocity monitoring. As highlighted in Sect. 3.1.1, the first of these was the binary central star of NGC 2346 (Méndez and Niemela 1981). However, the next definitive detection would be a long time coming, with the discovery of the double-degenerate system at the heart of Fleming 1 (Fg 1; Boffin et al. 2012b). This discovery highlights the strengths of the methodology, with some twenty spectroscopic observations taken randomly distributed throughout a six-month observing period revealing the 1.2 day orbital period, while previous photometric monitoring efforts failed to detect any signs of variability. A similar story can be told for the binary central star of MyCn 18, although its longer orbital period ($\sim$18 days) and, therefore, lower radial velocity semi-amplitude (11 km s^{-1} c.f. 88 km s^{-1} for Fg 1) meant that a higher-resolution spectrograph was needed to reveal the binary's orbital motion (Miszalski et al. 2018b).

Given that the progenitors of all PNe are necessarily (pre-)WDs with temperatures of several tens of thousands of degrees, another route to the discovery of binary central stars is through the presence of much cooler stellar signatures in their spectra or spectral energy distributions. These signatures can be revealed directly via spectroscopic observations or spectrophotometrically via the acquisition of precision multi-band photometry. The latter has been successfully employed by De Marco et al. (2013), Douchin et al. (2015), and Barker et al. (2018) to find approximately 30 systems whose spectral energy distributions are consistent with the presence of a cool companion. However, while this technique can provide an estimate of the temperature and spectral type of the companion, it does not reveal any information about the orbital period which always necessitates some form of time-resolved observation, be that spectroscopic (radial velocities) or photometric (light curves). As such, binary central stars revealed by this so-called "infrared excess" may be short-period post-CE systems or much wider, non-interacting systems which are simply close enough not to be resolved. Additionally, this technique relies upon the obtention of extremely precise photometry of the central stars in multiple photometric bands, across which the nebular structure and brightness may vary significantly. This greatly limits searches for spectrophotometric signs of infrared excess to faint and extended nebulae (De Marco et al. 2013; Douchin et al. 2015), which may be more likely to originate from single stars (Soker and Subag 2005).

3.2 Parameters of Known Systems

3.2.1 Period Distribution

As highlighted previously, for most of the sample of known close-binary central stars, only their orbital periods are known—a full list of which is presented in Table 3.1. The observed orbital periods are found to cluster between a few hours and a few days, with only a handful of systems found at longer periods. Various authors have remarked that the observed distribution is remarkably similar to that of the general post-CE population (e.g., Miszalski et al. 2009a), particularly when only considering systems with main sequence companions (Hillwig 2011; Nebot Gómez-Morán et al. 2011, and Fig. 3.10). However, given that the number of known post-CE central stars is now approaching a significant sample, we can now put this hypothesis to the test statistically—is the post-CE central star period distribution consistent with being randomly drawn from the same population as the general post-CE population? Isolating only post-CE central stars with main-sequence companions (by excluding those which show significant ellipsoidal modulation in their light curves, see Sect. 3.1.3.2, or which have confirmed degenerate or giant secondaries) results in a sample of 38 systems which can be compared to the white-dwarf-main-sequence (WDMS) binaries discovered using the Sloan Digital Sky Survey (SDSS, e.g., Rebassa-Mansergas et al. 2016, comprising 90 systems based on the list maintained at https://sdss-wdms. org/). A k-sample Anderson-Darling test calculated for the two complete samples

Table 3.1 A table of known post-common-envelope binary central stars of planetary nebulae

PN G	Common name	Orbital period (days)	Reference(s)
000.2 − 01.9	M 2-19	0.670	Miszalski et al. (2009a)
000.5 − 03.1a	MPA J1759−3007	0.503	Miszalski et al. (2009a)
000.6 − 01.3	Bl 3-15	0.270	Miszalski et al. (2009a)
000.9 − 03.3	PHR J1801−2947	0.316	Miszalski et al. (2009a)
001.2 − 02.6	PHR J1759−2915	1.104	Miszalski et al. (2009a)
001.8 − 02.0	PHR J1757−2824	0.799	Miszalski et al. (2009a)
[a]001.8 − 03.7	PHR J1804−2913	6.660	Miszalski et al. (2009a)
001.9 − 02.5	PPA J1759−2834	0.305	Miszalski et al. (2009a)
002.0 − 13.4	IC 4776	3.11	Miszalski et al. (2019b)
[a]003.1 − 02.1	PHR J1801−2718	0.322	Miszalski et al. (2009a)
[a]004.0 − 02.6	PHR J1804−2645	0.625	Miszalski et al. (2009a)
005.0 + 03.0	Pe 1-9	0.140	Miszalski et al. (2009a)
005.1 − 08.9	Hf 2-2	0.399	Hillwig et al. (2016a)
009.6 + 10.5	A 41	0.226	Bruch et al. (2001)
034.5 − 06.7	NGC 6778	0.15	Miszalski et al. (2011c)
049.4 + 02.4	Hen 2-428	0.176	Santander-García et al. (2015)
053.8 − 03.0	A 63	0.456	Afşar and Ibanoğlu (2008)
054.2 − 03.4	Necklace	1.16	Corradi et al. (2011b)
055.4 + 16.0	A 46	0.472	Afşar and Ibanoğlu (2008)
058.6 − 03.6	Nova Vul 2007	0.068	Rodríguez-Gil et al. (2010)
068.1 + 11.0	ETHOS 1	0.53	Miszalski et al. (2011a)
075.9 + 11.6	AMU 1	2.928	De Marco et al. (2015)
076.3 + 14.1	Pa 5	1.12	De Marco et al. (2015)
086.9 − 03.4	Ou 5	0.364	Corradi et al. (2014)
135.9 + 55.9	TS 01	0.163	Tovmassian et al. (2010)
136.3 + 05.5	HFG 1	0.583	Exter et al. (2005)
144.8 + 65.8	LTNF 1	2.29	Ferguson et al. (1999)
215.6 + 03.6	NGC 2346	16	Brown et al. (2019)
197.8 + 17.3	NGC 2392	1.90	Miszalski et al. (2019a)
220.3 − 53.9	NGC1360	142	Miszalski et al. (2018a)
221.8 − 04.2	PM 1-23	1.26[b]	Manick et al. (2015)
242.6 − 11.6	M 3-1	0.127	Jones et al. (2019)
253.5 + 10.7	K 1-2	0.676	Exter et al. (2003)
259.1 + 00.9	Hen 2-11	0.610	Jones et al. (2014)
283.9 + 09.7	DS 1	0.357	Hilditch et al. (1996)
290.5 + 07.9	Fg 1	1.195	Boffin et al. (2012b)
307.2 − 03.4	NGC 5189	4.04	Manick et al. (2015)
307.5 − 04.9	MyCn 18	18.15	Miszalski et al. (2018b)
329.0 + 01.9	Sp 1	2.91	Hillwig et al. (2016b)

(continued)

Table 3.1 (continued)

PN G	Common name	Orbital period (days)	Reference(s)
331.5 − 02.7	Hen 2-161	1	Jones et al. (2015)
332.5 − 16.9	HaTr 7	0.322	Hillwig et al. (2017)
335.2 − 03.6	HaTr 4	1.74	Hillwig et al. (2016a)
337.0 + 08.4	ESO 330-9	0.296	Hillwig et al. (2017)
338.1 − 08.3	NGC 6326	0.37	Miszalski et al. (2011c)
338.8 + 05.6	Hen 2-155	0.148	Jones et al. (2015)
341.6 + 13.7	NGC 6026	0.528	Hillwig et al. (2010)
349.3 − 01.1	NGC 6337	0.173	Hillwig et al. (2016b)
355.3 − 03.2	PPA J1747−3435	0.225	Miszalski et al. (2009a)
[a]355.6 − 02.3	PHR J1744−3355	8.234	Miszalski et al. (2009a)
358.8 + 04.0	Th 3-15	0.151	Soszyński et al. (2015)
355.7 − 03.0	H 1-33	1.128	Miszalski et al. (2009a)
354.5 − 03.9	Sab 41	0.297	Miszalski et al. (2009a)
357.0 − 04.4	PHR J1756−3342	0.266	Miszalski et al. (2009a)
357.1 − 05.3	BMP J1800−3407	0.145	Miszalski et al. (2009a)
357.6 − 03.3	H 2-29	0.244	Miszalski et al. (2009a)
358.7 − 03.0	K 6-34	0.39	Miszalski et al. (2009a)
359.1 − 02.3	M 3-16	0.574	Miszalski et al. (2009a)
359.5 − 01.2	JaSt 66	0.276	Miszalski et al. (2009a)

[a]Identified as a binary by Miszalski et al. (2009a) but since listed by Miszalski et al. (2011b) as uncertain and in need of confirmation
[b]Hajduk et al. (2010) erroneously identify a period half this value

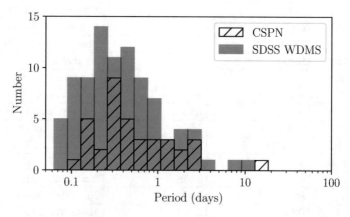

Fig. 3.10 The period distributions of WDMS binaries discovered using SDSS and of close binary CSPNe where the secondary is suspected to be a main-sequence star

shows that the null hypothesis can be rejected at approximately the 10% level—while not entirely conclusive, this could be taken to indicate that the two populations are not drawn from the same underlying distribution. However, this is likely to be a selection effect given that many of the known post-CE central stars have been discovered via targeted searches rather than uniform surveys. Furthermore, if the same test is repeated using post-CE central stars with suspected main sequence companions discovered by the OGLE survey only, then the null hypothesis cannot be rejected (significance level >50%).

3.2.2 Light and Radial Velocity Curve Modelling

Beyond the derivation of orbital periods from photometric monitoring, several systems have also been the subject of detailed studies revealing more information about their properties. These studies are, generally, based upon the modelling of observed light and radial velocity curves using a computer programme to create a synthetic binary star from which one can compute synthetic observables (model light and radial velocity curves). As one can imagine, the creation of such a synthetic model can involve a great deal of complexity with multiple possible (and valid) methodologies for determining the stellar morphologies (and discretizing their surfaces) and deriving the observed flux from each surface element (depending on local temperature, surface gravity, abundances, choice of model atmosphere, limb-darkening, gravity-brightening, etc.) as well as whether a given element, or some fraction thereof, is visible at a given time. For a full discussion of the process, the reader is referred to Prša (2019).

The codes used to model close binary stars vary on how they tackle each aspect of the problem, employing different methodologies or accounting for different physics—all of which can introduce additional uncertainties into the final models. The most prominent codes used to model binary stars are all based (even loosely) on the so-called Wilson-Devinney code (Wilson and Devinney 1971), which has been continuously expanded, improved and adapted (Wilson 1979, 2008; Wilson and Van Hamme 2014). One other code worth mentioning is the PHOEBE 2 code which has been used extensively throughout this book and presents several refinements over the Wilson-Devinney code (many of which are discussed in Prša et al. 2016; Horvat et al. 2018).

Beyond the limitations (or biases) of a given modelling code, it is also important to appreciate that the choice of fitting algorithm can also introduce significant issues. Given that fitting to binary observables is a highly non-linear problem, where the final light or radial velocity curve has a very complex dependency on the parameters of the modelled binary, simple minimization algorithms can often derive a combination of wrong (sometimes unphysical) parameters that seemingly provide a reasonable fit to the data. With access to modern supercomputing facilities, the preferred methodology to avoid such pitfalls is to employ a heuristic approach, like Markov Chain Monte Carlo sampling (Jones et al. 2019).

3.2.2.1   Masses, Radii and Temperatures

The principal parameters which one would wish to obtain from modelling are the masses, radii and temperatures of both stars, as well as the inclination of the binary orbital plane with respect to the observer. As most systems were discovered via photometric monitoring in a single photometric band, it is important to understand the limitations of such data in determining these parameters. In general, from a single monochromatic light curve of an eclipsing binary, the only parameters that can be constrained are the orbital inclination and radii relative to the orbital separation (absolute radii depend on the orbital separation which is a function of the stellar masses) and, to a lesser extent, the ratio of effective temperatures. For non-eclipsing binaries, almost no information can be derived from a single mono-chromatic light curve as the degeneracies between individual parameters are too large.

In order to constrain the masses of the components, radial velocity observations are required.[3] Double-lined radial velocity curves (i.e. those where the radial velocities of both stars can be measured) provide a direct measure of the mass ratio of the two stars, which when combined with the orbital inclination (usually determined by modelling of the light curve) can be used to derive the individual masses. However, it is important to note that irradiation complicates the mass measurements significantly. Depending on the choice of line or lines used to derive the radial velocity of the irradiated star, the semi-amplitude can be a (very) poor representation of dynamical velocity of the star (as the photocentre is often shifted towards the irradiating star). Thus the mass ratio derived is artificially reduced. However, modern modelling codes can go someway to combating this issue as they derive synthetic radial velocities by taking a photometrically-weighted integral of the surface radial velocities of individual surface elements (thus accounting for the shift in photocentre[4]; Prša et al. 2016; Horvat et al. 2019).

Multi-band light curves, or atmospheric modelling of spectroscopic data, are required in order to constrain the temperatures of the stars in a close-binary. Multi-band photometry can be used to constrain the stellar temperatures (as long as they are different) as the relative brightnesses of the two stars vary as a function of wavelength and therefore passband. Atmospheric modelling is generally underemployed in the modelling of binary central stars, but can provide strong constraints on the temperature and surface gravity of the hot star, which can then refine the simultaneous light and radial velocity curve modelling (Hillwig et al. 2017). Given the now well-established link between nebular and binary inclinations (see Sect. 3.3 and Hillwig et al. 2016b), the nebular morphology can also be used to provide rough constraints on the binary orbital inclination where necessary.

To date, only a handful of systems have been the subject of simultaneous light and radial velocity curve monitoring with those that result in strong constraints listed in

---

[3]Atmospheric modelling of spectra can derive the surface gravities, which in also be used to place constraints on masses when combined with radii measurements from light curve models.

[4]This is only an approximation as different emission/absorption lines can be produced in different regions of the star such that the photocentre may still not be an accurate representation.

Table 3.2. Note that a couple of other systems have also been investigated but, either due to the lack of radial velocity observations (see e.g. Bruch et al. 2001; Jones et al. 2014; Hillwig et al. 2016a) or their non-eclipsing natures (see e.g. Exter et al. 2003; Exter et al. 2005), could not be well-constrained. However, in spite of the relatively small number of systems that have well-constrained masses, radii and temperatures, there are several surprises to be found.

Focusing firstly on the parameters of the primaries, there appear to be a significant number of systems where the primary has a mass consistent with being a post-RGB star (Hillwig et al. 2017). While many models of CE interaction predict large numbers of systems to survive a CE on the RGB (in spite of the more massive and more tightly bound envelope), they are generally not predicted to form visible PNe as the post-RGB evolutionary times are thought to be longer than PN lifetimes (e.g. Nie et al. 2012). However, the models of Hall et al. (2013) show that it is possible for post-RGB CE core masses as low a 0.3 $M_\odot$ to form observable PNe. It is also interesting to note here that the best candidate post-RGB central star—that of ESO 330-9—is consistent with post-RGB tracks of the same mass but for ages much greater than would be expected for a PN (Hillwig et al. 2017). This perhaps highlights a fundamental issue with the application of such tracks (both post-RGB and post-AGB), in that due to our incomplete knowledge of the CE phase they are almost certainly not valid for post-CE systems (Miller Bertolami 2017). Therefore, it is not surprising that similar (or greater) discrepancies are found between such evolutionary tracks and other binary central stars (e.g. Santander-García et al. 2015; Jones et al. 2019).

In terms of the secondaries in these systems, those on the main sequence display discordant parameters with masses, temperatures and radii that are not consistent with models of isolated stars. In general, the secondaries display radii much greater than those expected for their masses as well as much greater temperatures. The increase in temperature can be well explained as a consequence of the high levels of irradiation in these systems, particularly if a significant fraction of the irradiating flux is absorbed and/or redistributed across the stellar surface (De Marco et al. 2008; Horvat et al. 2019). Similarly, irradiation may also contribute towards inflating the stellar radii, but it is likely that this is more a consequence of rapid accretion onto the secondary either just prior to or during the CE phase. This rapid accretion from the primary onto the secondary knocks the star out of thermal equilibrium resulting in the observed inflation. Low-mass stars (such as the observed secondaries) are either fully convective or have large convective zones meaning that the amount of material that must be accreted to cause this inflation is small, while the accretion rate must be significant (Prialnik and Livio 1985). However, recent models demonstrate that this required rate could easily be exceeded under the right circumstances (Chamandy et al. 2018), offering clear support for this accretion as the cause of the observed inflated secondaries. Further support is found in the morphologies and kinematical ages of the surrounding nebulae (see Sect. 3.3 and Jones et al. 2015), as well as the observed chemical contamination in the secondary of the Necklace (see Chap. 5 and Miszalski et al. 2013a).

In spite of the small numbers, one can also note that all of the main-sequence secondaries listed in Table 3.2 are rather low mass (K- or M-type). Even if one expands

Table 3.2 Close binary CSPNe with well-constrained masses, radii and temperatures derived from simultaneous light and radial velocity curve modelling. The horizontal line is a demarcation between systems with main sequence secondaries (upper) and double-degenerate systems (lower)

PN	Period (days)	M_{CS} ($M_\odot$)	R_{CS} ($R_\odot$)	T_{CS} (kK)	M_S ($M_\odot$)	R_S ($R_\odot$)	T_S (kK)	i (°)	References
Abell 46	0.47	0.51 ± 0.05	0.15 ± 0.02	49.5 ± 4.5	0.15 ± 0.02	0.46 ± 0.02	3.9 ± 0.4	80.3 ± 0.1	(a)
Abell 63	0.47	0.63 ± 0.05	0.35 ± 0.01	78 ± 3	0.29 ± 0.03	0.56 ± 0.02	6.1 ± 0.2	87.1 ± 0.2	(a)
Abell 65	1.00	0.56 ± 0.04	0.056 ± 0.008	110 ± 10	0.22 ± 0.04	0.41 ± 0.05	5.0 ± 1.0	61 ± 5	(b)
DS 1	0.36	0.63 ± 0.03	0.16 ± 0.01	77 ± 3	0.23 ± 0.01	0.40 ± 0.01	3.4 ± 1	62.5 ± 1.5	(c)
ESO 330-9	0.30	0.38–0.45	0.03–0.07	55–65	0.3–0.5	0.35–0.50	≤ 4.5	7–13	(d)
HaTr 7	0.32	0.50–0.56	0.13–0.18	90–100	0.14–0.20	0.3–0.4	≤ 5	45–50	(d)
Hen 2-155	0.15	0.62 ± 0.05	0.31 ± 0.02	90 ± 5	0.13 ± 0.02	0.30 ± 0.03	3.5 ± 0.5	68.8 ± 0.8	(e)
LTNF 1	2.29	0.70 ± 0.07	0.08 ± 0.01	105 ± 5	0.36 ± 0.07	0.72 ± 0.05	5.8 ± 0.3	84 ± 1	(f)
M 3-1	0.13	0.65^*	0.41 ± 0.02	48^{+17}_{-10}	0.17 ± 0.02	0.23 ± 0.02	5–12	75.5 ± 2	(g)
NGC 6337	0.17	0.56^*	0.045–0.085	115 ± 5	0.14–0.35	0.30–0.42	4.5 ± 0.5	17–23	(h)
Sp 1	2.91	0.52–0.60	0.20–0.35	80 ± 10	0.52–0.90	1.05–1.60	3.5–4.6	7–11	(h)
Hen 2-428	0.18	0.88 ± 0.13	0.68 ± 0.04	32.4 ± 5.2	0.88 ± 0.13	0.68 ± 0.04	30.9 ± 5.2	64.7 ± 1.4	(i)
NGC 6026	0.53	0.57 ± 0.05	1.06 ± 0.05	38 ± 3	0.57 ± 0.05	0.05 ± 0.01	146 ± 15	82 ± 2	(j)
TS 01	0.16	0.71–0.93	0.10–0.13	160–200	0.54^*	0.43 ± 0.02	57^*	45–75	(k)

*Fixed in the modelling

References: (a) Afşar and Ibanoğlu (2008), (b) Hillwig et al. (2015), (c) Hilditch et al. (1996), (d) Hillwig et al. (2017), (e) Jones et al. (2015), (f) Ferguson et al. (1999), (g) Jones et al. (2019), (h) Hillwig et al. (2016b), (i) Santander-García et al. (2015), (j) Hillwig et al. (2010), (k) Tovmassian et al. (2010)

the sample to include systems that do not have such well-constrained parameters, then there is still an absence of massive secondaries. The central star of NGC 2346 is seemingly the only exception, where the mass of the secondary is at least 3.5 $M_\odot$ (Brown et al. 2019). The properties of the general WDMS population show a very similar absence of early-type secondaries to that seen in post-CE central stars (Davis et al. 2010), strongly indicating that this is not due to small number statistics but in fact a real absence of such secondaries in post-CE systems (although it may well also be a consequence of the large difference in optical brightness between the two binary components preventing their detection; Parsons et al. 2016). This could perhaps be a consequence of the apparent dependency of the CE efficiency on the mass of the secondary, where the ejection efficiency has been found to be inversely proportional to secondary mass (Davis et al. 2012).

Using the parameters of the single-degenerate systems from Table 3.2, one may try to be more quantitative with regards their pre-CE parameters. Assuming the CE did not drastically alter the final core mass (i.e. cut the evolution too early), one can use an initial-final mass relation (such as that of El-Badry et al. 2018), to compute the initial mass of the AGB prior to the Roche-lobe overflow (RLOF). Assuming that the companion didn't accrete much mass and that the AGB didn't lose much mass prior to the RLOF, one can then estimate the initial masses of the binary components (as in, for example, Iaconi and De Marco 2019). It appears that, with only the exception of Sp 1 (where the uncertainties encompass a range of possible pre-CE mass ratios), all systems had primaries which were at least five times more massive than their companions before passing through the CE! Given that the mass ratio distribution of solar-like stars is more or less uniform (Moe and Di Stefano 2017), the fact that we only see such extreme pre-CE mass ratios is particularly interesting. Moreover, as we will see in Chap. 4, there are many post-mass transfer systems that have rather long orbital periods, of hundreds to several thousands of days—apparently having avoided a full CE episode. These have, on average, solar-like companions to the (pre-)white dwarf, that is, the initial mass ratio was relatively close to one—perhaps indicative that systems with more massive companions might be able to remove the primary's envelope more effectively, while low-mass companions necessitate a CE event resulting in very close orbits (Eggleton 2006).

3.2.3 Double-Degenerates

Only three double-degenerate (DD) central stars have been subjected to detailed modelling based on both photometric and spectroscopic observations, however two of those were found to present with total masses close to or greater than the Chandrasekhar mass and a time to merger shorter than the age of the Universe (see lower part of Table 3.2; Tovmassian et al. 2010; Santander-García et al. 2015). This is of particular interest as such systems are considered strong candidate Type Ia supernova (SN) progenitors (Maoz et al. 2014). However, neither of these systems (the central stars of Hen 2-428 and TS 01) have been robustly confirmed—the uncertainties on the total mass of TS 01 encompassing sub-Chandrasekhar values, while there have been doubts raised about the total mass of Hen 2-428 (Reindl et al. 2018). In spite of this, DD central stars of PNe would seem to represent a happy hunting ground for finding a bona-fide progenitor given that a significant fraction of SNe Ia are found to explode in circumstellar environments consistent with being inside a remnant PN (Tsebrenko and Soker 2015).

The DD fraction amongst the known binary central star population is also seemingly much higher than population synthesis models predict (Hillwig et al. 2010; Jones and Boffin 2017a). This conclusion is based on the assumption that all central stars found to display ellipsoidal variability (see Sect. 3.1.3.2) are DD systems—a seemingly reasonable proposition as every ellipsoidally-modulated central star system that has been subjected to further study has been found to be DD (Shimanskii et al. 2008; Tovmassian et al. 2010; Hillwig et al. 2010; Santander-García et al. 2015). If this assumption is valid then a very significant fraction of all close-binary CSPNe are DD systems. Of the 22 new binary central stars discovered via the OGLE survey (Miszalski et al. 2009a; Soszyński et al. 2015), six are ellipsoidal variables—some 27%. Furthermore, given that DD systems will generally only be detected by such surveys for photometric variability when one or both components are Roche-lobe filling, the true fraction is likely even greater (Boffin et al. 2012b; Jones and Boffin 2017a).

3.3   Relationship Between Central Stars and Their Host Nebulae

3.3.1   Axisymmetry, Polar Outflows and Filaments

As mentioned previously, the nebular material of post-CE PNe is believed to be comprised primarily of the ejected CE. Clearly, this means there should be strong relationships between the properties of the central binaries and their surrounding PNe. Given that the ejection of the CE is intrinsically aspherical (see Chap. 2 and, for example, Nordhaus and Blackman 2006), PNe around post-CE central stars should present aspherical morphologies. This is clearly borne out amongst the known population with some 30% presenting with canonical bipolar morphologies (possibly up to 60% when accounting for inclination effects; Miszalski et al. 2009b). Furthermore, in the handful of cases where inclinations of both binary orbital plane and nebular symmetry axis have been derived, the two are found to lie perpendicular—where the probability of finding such an alignment by chance is less that one in a million (Hillwig et al. 2016b). This is, to-date, the strongest statistical demonstration of the impact of central star binarity on the morphologies of PNe.

Beyond the prevalence of axisymmetric structures in PNe with close-binary central stars, Miszalski et al. (2009b) also found a high occurrence rate of low-ionisation structures in the form of knots and filaments indicative of a binary origin. This is of particular interest in the context of nebular chemistry, wherein several PNe with close

Table 3.3 The kinematical ages of PNe with polar outflows and binary central stars. Due to the large uncertainties on the distances to these nebulae, all ages are given per unit distance

PN	Age of central nebula (yrs kpc^{-1})	Age of polar outflow (yrs kpc^{-1})	Reference
Abell 63	3500 ± 200	5200 ± 1200	Mitchell et al. (2007)
ETHOS 1	900 ± 100	1750 ± 250	Miszalski et al. (2011a)
Necklace	1100 ± 100	2350 ± 450	Corradi et al. (2011b)
Fg 1	~ 2000	~ 2500–7000	Lopez et al. (1993), Boffin et al. (2012b)
NGC 6337	~ 9200	~ 1000	García-Díaz et al. (2009)
NGC 6778	~ 1700	~ 650	Guerrero and Miranda (2012)

binary central stars have been shown to present with abundance patterns consistent with the presence of a second colder, higher-metallicity gas phase similar to those exhibited by the knotty structures around the born-again PNe Abell 30 and Abell 58 (Wesson et al. 2005, 2008). For further discussion of this, please see Chap. 5.

Miszalski et al. (2009b) also identified a prevalence of jet-like structures amongst the population of PNe known to host binary central stars. Jets are relatively easy to understand in a binary scenario, with such structures appearing as a natural consequence of mass transfer between the two stars (Tocknell et al. 2014). However, spatio-kinematical studies of these jets have revealed that in most cases they are found to be older than the central regions of their respective nebulae, indicative that they were not ejected simultaneously (see Table 3.3). The likely explanation for this is that the jets are launched during a period of intense accretion onto the secondary just prior to entering into the CE, which is ejected some time later going on to form the central nebular regions (Tocknell et al. 2014). This chronology is also supported by the apparent precession period of the jets observed in Fleming 1, which is consistent with an orbital period much greater than the current post-CE period (Boffin et al. 2012b). Apparent exceptions are found in NGC 6337 and NGC 6778, where the jets are thought to have formed after the ejection of the CE—in these cases the jets may, in fact, originate from accretion onto the primary (rather than onto the secondary, Soker and Livio 1994) or by chaotic accretion onto the companion at the termination of the CE (Soker 2019).

It is important to note that all the morphological traits identified by Miszalski et al. (2009b), based on the unbiased sample of binary central stars discovered by the OGLE survey, as being typical of binarity have since been successfully used by targeted surveys in increasing the binary discovery rate (see Sect. 3.1.2).

3.3.2 Nebular Masses

The detailed study of the archetypal post-CE PN Abell 63 by Bell et al. (1994) was the first to reveal the tendency of post-CE PNe to present with rather low ionised masses. This finding has since been confirmed for a multitude of objects where, even though there is significant uncertainty over the distances, electron densities and electron temperatures of the nebulae (the latter two being due to the apparent presence of multiphase gas with differing temperatures and densities, see Chap. 5), in many cases the nebular ionised masses are several orders of magnitude lower than those found in the general PN population (Liu et al. 2006; Corradi et al. 2011a, 2015a). Here, it is particularly interesting to note that the general PN population already displays a tendency towards lower ionised masses than one might expect, considering that they should comprise a large fraction of the AGB's envelope and most of this material should be ionised (Pottasch 1980; Buckley and Schneider 1995). However, hydrodynamic models have shown that the majority of this ionised material should be found in large, low-density, low-luminosity haloes which would likely go undetected (Villaver et al. 2002), thus explaining the generally underestimated total ionised masses (in the, perhaps extreme, case of Abell 36, the mass excluding the halo is roughly three orders of magnitude lower than the true total! McCullough et al. 2001).

Given that post-CE PNe are expected to represent the complete ejection of the primary's envelope (rather than its slower exhaustion via a dense AGB wind), these PNe should, in general, be more massive than those produced via a single star evolution (Santander-García et al. 2019). One possibility is that much of the mass in these systems is not ionised, but rather is found in a molecular or neutral state. However, preliminary observations seem to indicate that this is not the case, with a sample of post-CE PNe being found to present no significant molecular component (Santander-García et al. 2019). Further, systematic study of a significant number of post-CE PNe, including the determination of ionised, molecular and neutral masses (as well as their distributions), is required before any strong conclusions regarding post-CE masses can be drawn. However, this may be a strong indication that the entire CE is not ejected in one event rather in multiple episodes as in the grazing envelope evolution (Soker 2015, 2017), an idea perhaps supported by the apparent inability of hydrodynamic models to completely eject the CE (see e.g., Passy et al. 2012a; Ohlmann et al. 2016; Oomen et al. 2018; Iaconi et al. 2018, and Chap. 2).

Chapter 4
Long Period Central Stars of Planetary Nebulae

4.1 Making the Case

It is now unambiguously clear that the majority of solar-like stars are members of binary or multiple systems (Duquennoy and Mayor 1991; Halbwachs et al. 2003; Raghavan et al. 2010; Whitworth and Lomax 2015; Boffin 2019). Most recently, Fuhrmann et al. (2017) showed that 58% of F- and G-type Population I stars are in binary or multiple systems. These studies also showed that the binarity/multiplicity rate increases with the mass of the primary star. The orbital period distribution is well approximated by a Gaussian centred on $\log P(\text{days}) = 5.03$ and a standard deviation $\sigma_{\log P} = 2.28$, which, assuming a total mass of ~ 1.5 $M_{\odot}$, translates to a mean separation between objects of ~ 50 au. As seen before (Chap. 2), systems with periods above, say, 1,000 days (but see below) will not experience a common envelope evolution which would lead to a final orbital period below a few days, and we may therefore expect that a large fraction ($\sim 45\%$) of all solar-like stars will end up in intermediate- and long-period binary systems, comprising a white dwarf (WD) and a companion, which could be of any kind, between an M dwarf and an A/F main-sequence star (as the mass-ratio distribution of solar-like stars is basically uniform; Boffin et al. 1993; Raghavan et al. 2010; Boffin 2010; Boffin and Pourbaix 2018). And indeed, Van der Swaelmen et al. (2017) found a fraction of 22% of post-mass-transfer systems (i.e. with a white dwarf companion) among cluster binary giants, while Murphy et al. (2018) found that among intermediate-mass single-pulsator binaries with orbital periods between 100 and 1500 d and having a mass ratio larger than 0.1, a large fraction of the companions ($21 \pm 6\%$) are white dwarfs. As some of these white dwarfs possibly went through the planetary nebula (PN) phase, we may thus expect to find a few central stars of PNe that comprise long-period binaries—with a handful being discovered in recent years.

The discovery of long-period central stars is essential as they allow to study mass transfer by stellar wind (Boffin 2015) and via wind-Roche lobe overflow (Nagae et al. 2004; Jahanara et al. 2005). Hydrodynamic simulations (see also Theuns et al. 1996;

© The Author(s), under exclusive license to Springer Nature Switzerland AG 2019
H. M. J. Boffin and D. Jones, *The Importance of Binaries in the Formation and Evolution of Planetary Nebulae*, SpringerBriefs in Astronomy,
https://doi.org/10.1007/978-3-030-25059-1_4

Edgar et al. 2008) have shown that this kind of mass transfer can also affect the shape of the nebula, even up to very long periods. Thus, Maercker et al. (2012) showed a spiral structure in the material around the asymptotic giant branch (AGB) star R Sculptoris (as expected from numerical simulations) and infer an orbital period of about 350 years (1.3×10^5 d). When the AGB will evolve into a hot WD that will ionise this nebula, there is no reason to believe that the resulting PN will be spherically symmetric (see also, Ramstedt et al. 2017)! More recently, Brunner et al. (2019) found a thin, irregular, and elliptical detached molecular shell around the AGB star TX Piscium, whose origin is possibly related to the presence of a low-mass companion.

4.2   A Large Family

To understand the importance of such wide binaries and their link with other kinds of binaries, it is useful to consider again the fundamentals of stellar evolution of low- and intermediate-mass stars (see also Sect. 1.2). As the star leaves the main-sequence, it will evolve along the red giant branch (RGB), then after a core-Helium burning phase, it will ascend the AGB, where it reaches its highest luminosity and radius, and undergoes heavy mass-loss. The evolution of the radius of the star will depend on its initial mass, as stars less massive than about 2 $M_\odot$ will reach rather large luminosities and radii on the RGB, compared with more massive stars, while the largest radius reached on the AGB increases with initial mass. This is illustrated in Fig. 4.1, which shows the radius evolution for stars of various initial masses, based on the models of Hurley et al. (2000). This indicates that stars of 1, 1.5, 2, 3, and 4 $M_\odot$ will reach a maximum RGB radius of, resp., 186, 155, 28, 46, and 76 $R_\odot$, and on the AGB, of 221, 306, 391, 614, 855 $R_\odot$. The actual values may vary somewhat (and be a function of, for example, metallicity) given that these are based on stellar models, but we are only interested here in the orders of magnitude. When such stars have binary companions, these maximum radii will define the minimum separations that are needed to avoid the star to fill its Roche lobe, thereby terminating its evolution. As such, a low-mass star is far more likely to fill its Roche lobe while on the RGB (as the difference in maximum radius on the RGB to that on the AGB is rather small). Stars more massive than 2.5 $M_\odot$, on the other hand, could experience Roche lobe overflow on either the RGB or AGB depending on the initial binary separation (which, given the observed main sequence period distribution means that mass transfer on the AGB is most likely).

We plot in Fig. 4.2 the Roche lobe radius at periastron of two primary stars of different masses in a binary system with a mass ratio[1] of 2, as a function of the orbital period, and compare this to the maximum radius reached on the tip of the RGB or the AGB. This provides a very rough idea of the orbital period at which Roche-lobe overflow (RLOF) would take place. Thus, for a 1.5 $M_\odot$ primary star, in a $q = 2$ binary system, RLOF would happen on the RGB for orbital periods, P, below 500 days ($e = 0$) or 1,550 d ($e = 0.5$), and on the AGB for, resp., $P = 1,500$ d and

---

[1] Here the mass ratio is defined as the mass of the more massive and most evolved star over the mass of its companion.

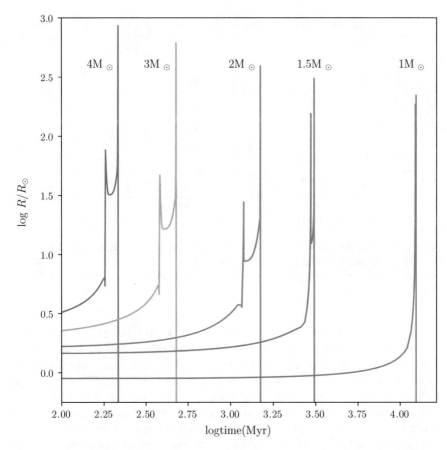

Fig. 4.1 The evolution of stellar radius as a function of age, in logarithm scale for various initial masses, based on the models of Hurley et al. (2000). The more massive a star, the more quickly it reaches the RGB and the AGB. The radius reached on the tip of the AGB increases with the initial mass, while it also appears that low-mass stars reach a much larger radius on the tip of the RGB. In a binary system, this will define when a star will fill its Roche lobe and terminates the evolution

$P = 4,500$ d. For a 3 $M_\odot$, only very short orbital periods would lead to a RLOF on the RGB, while on the AGB this could happen for periods below, resp., 2,800 and 8,000 days. These are only very rough estimates, as physical processes such as tidal circularisation, mass loss and accretion, as well as angular momentum loss will lead to very different values. For example, the BSE code (Hurley et al. 2002), used with the default parameters, shows that for an $e = 0.3$ system with a 3 $M_\odot$ primary star, in a $q = 2$ binary system, common envelope evolution happens for orbital periods below $P = 3,000$ d, stable RLOF happens until $P = 4,000$ d, while only wind mass transfer takes place beyond $P = 4,700$ d. As we are still far from completely understanding the mass transfer processes involved, these values should only be considered as very indicative. At the very least, they allow us to define some families of stars, which have links with the wide-binaries CSPNe. A useful presentation of these is available in Van Winckel et al. (2010).

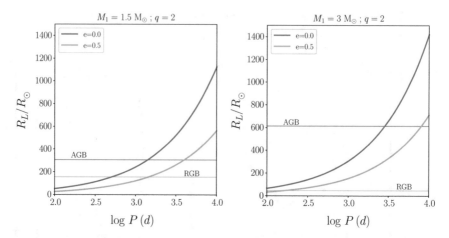

Fig. 4.2 The relation between the Roche-lobe radius at periastron of the primary of 1.5 $M_\odot$ (left) and 3 $M_\odot$ (right) and the orbital period, in logarithmic scale, for a mass ratio of 2, and two values of the eccentricities, $e = 0$ and 0.5. The maximum radius reached by the primary on the RGB and AGB is also indicated

As a side note, we could also use the value of $P = 3{,}000$ d shown above to estimate how many post-CE systems we should expect inside PNe, if binarity is not a factor influencing the *formation* of PNe. Given the period distribution of solar-like stars, we can thus expect to have about 24% of all binaries to lead to a CE. Assuming a 58% binary fraction, this leads to a total fraction of 14% of expected post-CE binary systems among solar-like stars. This is a very conservative number, however, which critically depends on the upper value of the period where a CE could occur. If this value is decreased to 1,000 days, say, the fraction becomes 10%. One can also make the argument that this number is an over-estimate as it includes all systems with orbital periods below the threshold, including systems that may undergo a CE too early to result in a PN (for example, if the orbital separation is particularly small, the CE could occur while on the main sequence) or systems which merge during the CE. As shown by Fig. 4.2, the possible range of orbital periods that would lead to a RLOF on the AGB is rather limited, especially for lower initial masses. Moreover, as shown in Sect. 3.2.2.1, post-CE close binaries seem to be characterised by a initial mass ratio (mass of the WD progenitor to its companion mass) which is very high— between 6 and 18 for most of the objects where this can be estimated. Given that the mass ratio distribution of solar-like stars is more or less uniform, this means that one would have to multiply the above numbers by a factor of $\sim$0.2! The fact that the observed *lower limit* fraction of post-CE PNe is 12–21% (Miszalski et al. 2009a), while the rough *upper limit* we derive is more like 2–5%, probably indicates that observable PNe more easily derive from a binary system than from a single star.

4.2.1 Symbiotic Stars

A symbiotic star is a binary system composed of a red giant (RGB or AGB) with a hot companion, most often a white dwarf, although there are also cases with a main-sequence or even with a neutron star companion (Mikołajewska 2003, 2012). The giant is transferring mass to its companion via stellar wind or (stable) Roche lobe overflow, but appear otherwise normal (Boffin 2014). Most of the symbiotic stars with measured orbital parameters have orbital periods between 200 and 1,000 days, with the longest reported being 44 years, but longer ones must exist. Some of the giants in symbiotic systems appear s-process enriched (Smith et al. 1996), which indicates that a previous episode of mass transfer (when the progenitor of the WD was on the AGB) took place. Hence, those symbiotic stars that contain a WD are experiencing their second phase of mass transfer and will possibly lead to double-degenerate systems. They are the clear demonstration (among with the other families described below) that mass transfer from a giant can be stable. This may imply that some additional mass loss is required compared to what is usually assumed for single stars (Blind et al. 2011).

4.2.2 Post-RGB Stars

Should the initial orbital period be small enough that the star will fill its Roche lobe while on the red giant branch, the evolution of the star will be prematurely terminated. If the mass ratio is large enough, a common envelope phase also occurs, leading to the formation of a close binary with a low-mass He-core WD or a core-He burning subdwarf B star component (perhaps leading to the formation of a post-RGB PN; Hall et al. 2013; Hillwig et al. 2017). In some cases, the stars will merge, forming a more massive single star. If the RLOF is stable, however, the RGB phase can be terminated, with the He core evolving to higher temperatures at near-constant luminosity, through the post-RGB phase. As we have seen above, this is most likely for systems containing primary stars with masses in the range $1–2.5\,M_\odot$. Such dusty post-RGB stars have only been discovered very recently, in the Magellanic Clouds, as the low-luminosity cousins to post-AGB stars (Kamath et al. 2016).

A system that may be closely related is IP Eri, which contains a K0 subgiant and an He WD in a 1071-day orbital period with an eccentricity of 0.25 (Merle et al. 2014), and whose existence requires again extreme mass-loss on the RGB. Further related objects are sdB wide binaries, which are thought to form only through stable Roche-lobe overflow of the sdB progenitor near the tip of the red giant branch (Vos et al. 2017, 2019). Quite interestingly, these last authors find a strong correlation between orbital period and the mass ratio.

4.2.3 Post-AGB Stars

When mass loss (either via stellar wind, or through mass transfer in a binary system) has reduced the stellar envelope to a very small amount, the star leaves the AGB. During the short time that it is not hot enough to ionise the material ejected during the AGB, the star is in a transitory phase known as post-AGB. Long-term monitoring have discovered a large sub-sample of binary post-AGBs among those post-AGBs that harbour a dusty circumbinary disc (Ertel et al. 2019), as indicated by their infrared excess. A very recent analysis of all post-AGB orbits (33 in total; Oomen et al. 2018) showed that their orbital periods are generally between 100 and 3,000 days, and a large fraction have non-negligible eccentricities. The companion masses are found to be broadly distributed around 1.09 ± 0.62 $M_\odot$. Based on their properties, one can deduce that the *current* Roche-lobe radii of the primaries are smaller than the radii reached by stars during the AGB phase, indicating that either they are the result of some stable RLOF or that some unknown processes led to a decrease in the separation of the system. The fact that most of these systems have non-circular orbits shows that perhaps tidal effects are not as strong as generally thought, although the presence of circumbinary discs would allow some eccentricity-pumping mechanisms to come in to play. Another possibility is that enhanced mass-loss at periastron passages, as is expected in the grazing envelope evolution, can compensate for the circularisation effect of the tidal interaction (Kashi and Soker 2018).

These disc post-AGB stars will evolve toward a white dwarf, but as there is no material surrounding them, except for the circumbinary disc, they will probably not become a planetary nebula (Lagadec 2018). Instead, the system will have all the properties of Barium stars and related objects (see below). However, the circumbinary discs in these systems are O-rich, which would indicate that the primary did not go through many thermal pulses, and are generally not s-process enriched (but see van Aarle et al. 2013). It is thus not sure that the companion will indeed appear as a Barium star. Instead, they may be the progeny of blue straggler stars as seen in open clusters (Boffin 2015).

In any case, post-AGB binaries highlight that disc formation must be a mainstream process in the late evolution of a very significant binary population (Van Winckel et al. 2010), which can drive the shaping of bipolar nebulae. Some of these systems also show clear signs of mass transfer, including an accretion disc around the primary and jets (Bollen et al. 2017), making their study even more interesting. See Chap. 7 for further discussion.

4.2.4 Barium Stars

Barium stars are G- and K-giants showing the clear signature of Carbon and s-process element enrichment in their spectra. Since their initial discovery, associated classes have been found among dwarf stars, more evolved stars (S stars) as well as metal-poor stars (CH and CEMP-s stars). It is now well established that Barium stars are binary stars, where the unseen companion is a white dwarf and the origin of the enrichment is due to a previous mass transfer episode while the WD was on the AGB (Boffin and Jorissen 1988; McClure and Woodsworth 1990; Boffin and Zacs 1994). Recent studies (Van der Swaelmen et al. 2017; Escorza et al. 2019; Jorissen et al. 2019) have provided the orbital properties for a very large sample of these stars. The Barium stars with strongest s-process over-abundances are observed in the period range 200–10,000 days, whereas mild Ba stars span the range 100–20,000 days. Almost all Barium stars with periods shorter than 1,000 d are in circular systems, while wider systems are all eccentric. The companion masses are compatible with the masses of WDs and there is a strong correlation found between s-process abundance and the inferred initial mass ratio, implying that strong Barium stars originate from systems with a mass ratio[2] above ∼1.5.

4.3 Long-Period Binaries in Planetary Nebulae

To discover wide binaries in PNe is not an easy task. There are basically two avenues to do so: a dedicated radial-velocity campaign or looking for the spectral/photometric signatures of cool companions. The second methodology comprises looking for central stars that either appear too red to be consistent with a hot ionising central star, or which show excesses in the near-infrared part of their spectral energy distributions consistent with the contribution of a cool companion. This latter method, however, is prone to error and provides no information on orbital separation (or even definitive proof of association), just as easily recovering some post-CE binaries (Barker et al. 2018), wide binaries in which no physical interaction can be expected (Ciardullo et al. 1999), and chance alignments (Jones and Boffin 2017a, b; Boffin et al. 2018).

Looking at the stars mentioned earlier, the binary systems that avoided the common envelope phase will have orbital periods in the range from 100 days to several years. The ensuing radial-velocity variations will thus be relatively small, of the order of a few to 10–20 km/s. Thus, such systems will only be discovered with systematic long-term monitoring using high-resolution spectroscopy. As of now, only a few such wide binaries could be found (Table 4.1), but this is most likely only the tip of the iceberg.

Most discoveries were rather recent as previous attempts proved unsuccessful, although the hopes were quite high. From a radial velocity survey, De Marco et al. (2004) found that 10 out of 11 CSPNe had variable radial velocities (RVs), suggesting that a significant binary population may be present. However, as far as we know, of these targets only one was confirmed as a binary, proving the difficulty of such observations. Other surveys were presented in Sorensen and Pollacco (2004) and Afšar and Bond (2005), but again no strictly periodic binaries were found, which could be due to a combination of intrinsic wind variability (Boffin et al. 2012a), but also to sparse sampling of the orbit and too large errors in deriving the RVs. Of the 13 possible binaries found by Sorensen and Pollacco (2004), only two (NGC 1514 and

---

[2]Defined as the mass of the WD progenitor to the mass of the current Barium star.

Table 4.1 Wide-binary central stars of planetary nebulae with known orbital periods

PN G	Common name	Orbital period (days)	Eccentricity	Refs.
052.7 + 50.7	BD + 33°2642	1105 ± 24	0.0[c]	[1]
165.5 − 15.2	NGC 1514	3306 ± 60	0.46 ± 0.11	[2]
220.3 − 53.9	NGC 1360	141.6 ± 0.8	0.0[c]	[3]
339.9 + 88.4	LoTr 5	2717 ± 63	0.26 ± 0.02	[4]
[a]	SMP LMC 88	~4900[b]	–	[5]

[a]In the Large Magellanic Cloud, no PN G number
[b]Inferred from photometric variability
[c]Fixed
References: [1] Van Winckel et al. (2014); [2] Jones et al. (2017); [3] Miszalski et al. (2018a); [4] Aller et al. (2018); [5] Iłkiewicz et al. (2018)

NGC 2346) have been confirmed.[3] Similarly, of the 12 apparently variable CSPNe of Afšar and Bond (2005), only two (NGC 1360 and NGC 2392) have since been confirmed. There is a clear need to perform higher precision, more systematic monitoring of the radial velocities of CSPNe. The few high-precision studies undertaken thus far have borne rather positive results.

Van Winckel et al. (2014) reported the long orbit of BD+33°2642 using the HERMES spectrograph on the Mercator telescope, and showed that the CSPN in LoTr 5 also displayed radial velocity variability consistent with a long-period orbit. The orbit of the latter was provided by Jones et al. (2017), who also discovered a long-period binary in NGC 1514—a PN known to host remarkable mid-infrared rings (Fig. 4.3). Aller et al. (2018) later provided a refined period for LoTr 5, also examining the possibility of a third component to the system. Miszalski et al. (2018a, b) found also a binary inside the PN NGC 1360, with a relatively smaller period (but still in line with what we have seen in this chapter) of about 142 days. They find that the CSPN of NGC 1360 is possibly a double-degenerate system.[4] More recently, Iłkiewicz et al. (2018) showed the presence of a yellow symbiotic binary inside the PN SMP LMC 88. If due to the orbital motion, the photometric variability would indicate an orbital period of the order of 4,900 days.

The orbital properties of these systems indicate that they are similar to those of post-AGB systems (indeed, BD+33°2642 presents abundance anomalies similar to those found in post-AGB binaries with circumbinary discs; Napiwotzki et al. 1994)— in apparent contradiction with the idea that post-AGB stars cannot produce PNe—but also to Barium stars (the secondary of LoTr 5 is, in fact, a barium star), so that they may also provide a link to these. This is confirmed by the discovery that several cool companions to CSPNe are enriched in s-process elements (Tyndall et al. 2013).

---

[3]They also identified the bright star slightly offset from projected centre of Sh 2-71 as presenting radial velocity variability consistent with binarity, a finding which was since confirmed by Močnik et al. (2015). However, there is considerable debate over the identity of the central star, with a fainter, bluer star found much closer to the projected nebular centre seemingly a stronger candidate (Frew and Parker 2007). See Jones et al. (2019) for further discussion.

[4]Quite interestingly, this system appears to have a very low inclination on the plane of the sky. If similar systems exist with a larger inclinations, the radial variations expected would be easily detectable, being a factor 1.7–2 times larger than observed in NGC 1360.

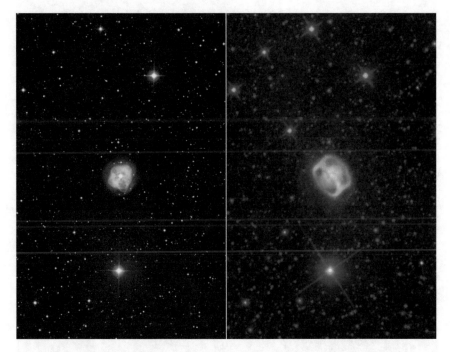

Fig. 4.3 The PN NGC 1514, known to host a long-period binary. The visible-light view on the left is from the DSS; the view on the right shows the object in infrared light, as seen by NASA's Wide-field Infrared Survey Explorer. Image credit: NASA/JPL-Caltech/UCLA/Digitized Sky Survey/STScI (see also Ressler et al. 2010)

Indeed, another way to discover potentially wide binaries inside PNe is to look for CSPNe with composite spectra or colours which are too red to be compatible with a hot ionising white dwarf (Bond et al. 1993; Bond 2000). One of the first class of such objects are the so-called *Abell 35-type* PNe in which a cool, rapidly rotating optically-bright star has a hot companion revealed by UV spectra. The rapid rotation is unmasked thanks to photometric variability on time scales of days, most likely due to the presence of spots on an active star. The first members of this group were Abell 35, LoTr 1 and LoTr 5—the latter one having now an orbit established. Somewhat ironically, Abell 35 is no longer a member of the Abell 35-type class, having been shown to be a Strömgren sphere around an evolved post-EHB star rather than a PN (Ziegler et al. 2012) and is thus not to be considered further here. However, some other members have been added recently: WeBo 1 (Bond et al. 2003; Siegel et al. 2012), A 70 (Miszalski et al. 2012), Hen 2-39 (Miszalski et al. 2013b; Löbling et al. 2019) and (potentially) Me 1-1 (which may, in fact, be a symbiotic star rather than a PN; Pereira et al. 2008).[5]

Interestingly, while LoTr 5, WeBo 1, A 70 and Hen 2-39 show a clear enhancement of s-process elements and sometimes of carbon, making them direct progenitors of

---

[5]Similarly, PC 11 is also a possible candidate but is rather more likely to be a symbiotic star (Pereira et al. 2010).

Table 4.2 The A35-like PNe

PN G	Common name	Barium enriched?	P_{rot} (days)	Refs.
038.1 − 25.4	A 70	y	2.061	[1, 2]
052.5 − 02.9	Me 1-1	n	–	[3]
135.6 + 01.0	WeBo 1	y	4.7	[4]
228.2 − 22.1	LoTr 1	n	6.4	[5]
283.8 − 04.2	Hen 2-39	y	5.46	[6, 7]
339.9 + 88.4	LoTr 5	y	5.95	[8]

References: [1] Miszalski et al. (2012), [2] Bond and Ciardullo (2018), [3] Pereira et al. (2008), [4] Bond et al. (2003), [5] Tyndall et al. (2013), [6] Miszalski et al. (2013b), [7] Löbling et al. (2019), [8] Aller et al. (2018)

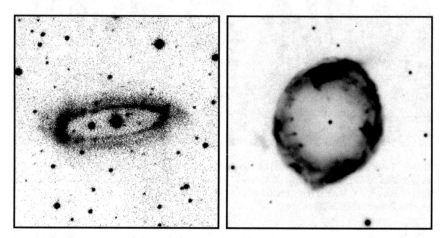

Fig. 4.4 Hα+[N II] imagery of Ba star PNe WeBo 1 (left) and A 70 (right) highlighting their ring-like morphologies. Data originally presented in Tyndall et al. (2013)

Barium stars, LoTr 1, on the other hand, doesn't appear enriched (Tyndall et al. 2013). As mentioned above, all the cool components inside these PNe are also rapidly rotating (for giant stars; see Table 4.2), which is an indication of having accreted material from the hot CSPN. These systems are thus Rosetta stones for the study of the mass transfer at the origin of Barium stars. The level of the s-process enhancement and the observed abundance pattern implies that mass transfer from a companion with an extremely high enrichment of AGB nucleosynthesis products is needed. Moreover, the inferred ratio of accreted to ejected mass indicates a quasi-conservative mass transfer. Thus wind-RLOF is preferred, while a wide binary involving Bondi–Hoyle–Lyttleton accretion can be ruled out as can a CE evolution (Löbling et al. 2019).

Finally, we note that all the PNe with Ba-rich central stars also present with ring-like waists and possibly extended lobes, consistent with bipolar morphologies (see Fig. 4.4 and Tyndall et al. 2013). The nebula of LoTr 1 consists, on the other hand, of two slightly elongated shells, with ages of 17,000 and 35,000 year, respectively, and with different orientations. This may be an indication of a difference in the mass-transfer history of this system with respect to the Ba star PNe.

Chapter 5
Chemical Evidence of Mass Transfer

We have shown in previous chapters how binarity affects the morphology and the kinematics of planetary nebulae (PNe). We have also seen that some PNe play host to wide binaries where one of the components is a star enriched in Carbon and/or s-process elements. In this chapter, we show that there is also a clear impact on the chemical composition of the central star and on its host nebula, for those binaries that went through the common envelope (CE) phase.

5.1 Accretion Prior to the Common Envelope

If the existence of Barium central stars of PNe (Ba CSPNe, see Chap. 4) constitutes firm evidence for mass transfer and accretion in PNe, close binary central stars (see Chap. 3) show no evidence for rapid variability (flickering) or spectroscopic features that could be attributed to *ongoing* accretion (except perhaps in the central star of NGC 2392; Miszalski et al. 2019a). On the other hand, the presence of outflows and/or jets surrounding several systems (e.g., Miszalski et al. 2011a), possibly launched from an accretion disc that is no longer present, may be indirect evidence for accretion, either prior to the common envelope phase via wind accretion from the AGB primary, at the onset of the CE phase or perhaps even after the CE phase. The jets could also be formed during the CE phase (Soker 2019), or towards the end of it and could then also help in ejecting the envelope (Chamandy et al. 2018; Shiber et al. 2019).

Kinematical observations of some jets around post-CE nebulae indicate, however, that the jets were probably ejected before their central nebulae (Mitchell et al. 2007; Corradi et al. 2011b; Miszalski et al. 2011a; Boffin et al. 2012b). Another fundamental clue comes from point-symmetric outflows of PNe. Simulations can recreate these complex outflows with a precessing accretion disc, around the secondary, which is launching jets (e.g. Cliffe et al. 1995; Raga et al. 2009). While there are multiple

© The Author(s), under exclusive license to Springer Nature Switzerland AG 2019
H. M. J. Boffin and D. Jones, *The Importance of Binaries in the Formation and Evolution of Planetary Nebulae*, SpringerBriefs in Astronomy,
https://doi.org/10.1007/978-3-030-25059-1_5

Fig. 5.1 FORS2 image of the PN Fleming 1, with its impressive precessing jets. Credit: ESO/H. Boffin

examples of such PNe, none were known to have a binary nucleus until the landmark discovery of a post-CE binary nucleus with a period of 1.2 days in the archetype of this class Fleming 1 (Boffin et al. 2012b, see Fig. 5.1). Fleming 1 possesses a spectacular set of bipolar jets spanning about 2.8 parsecs from tip to tip and delineated by a symmetric configuration of high speed knots. These follow a curved path, where each opposing pair of knots can be connected by lines that intersect precisely with the position of the central star. The outermost knots are elongated and were probably ejected about 16,000 years ago, whereas the innermost ones were possibly ejected some 10,000 years later. On the other hand, the innermost nebula is about 5,000 years old, assuming a constant expansion rate. Because of their appearance, suggestive of episodic ejections produced by a precessing source, Fleming 1 became the archetype of a morphological sub-class of PNe named after their bipolar, rotating, episodic jets (BRETs). The additional presence of a faint envelope detected in deep FORS2 narrow-band images adds further credence that we are witnessing the effect of an accretion disc around the white dwarf's companion star, that existed prior to the CE phase. This indicates that either wind accretion or some some stable RLOF took place in an earlier phase in the evolution of this system.

As compelling as it is, this is still indirect evidence. The clearest proof of accretion would be a polluted main-sequence companion with an atmosphere strongly enriched by accreted material. This is exactly what Miszalski et al. (2013a) found: the compan-ion to the post-CE central star of the Necklace planetary nebula (PN G054.2−03.4, Fig. 5.2) is enriched in Carbon. These authors found that to reproduce the observed Carbon enhancement, the companion needs to accrete between 0.03 and 0.35 $M_\odot$, depending on its mass, in a binary system with initial orbital period between 500 and 2 000 days. The current period of the system, 1.16 d, clearly proves that the system underwent a common-envelope phase. It is unclear if the accretion took place prior or during the CE: many simulations of the spiral-in phase (Passy et al. 2012b) predict a negligible amount of mass is accreted onto the secondary ($\sim 10^{-3}\ M_\odot$), although

Fig. 5.2 Image of the 'Necklace' Nebula, discovered by the IPHAS survey (Credit: Romano L. M. Corradi, IPHAS)

more recent simulations seem to imply that in some cases super-Eddington accretion could take place (Chamandy et al. 2018). However, wind accretion would also be able to form an accretion disc around the companion (Theuns et al. 1996). Since the jets of the Necklace are also observed to be older than the main PN (Corradi et al. 2011b), they were probably launched from such a disc. The initial period estimated for the system is typical of symbiotic stars, where substantial wind accretion onto companions is known to occur. The binary CSPN inside the Necklace nebula is most probably the progenitor of dwarf Carbon stars that have very short orbital periods, of which very few are known (Margon et al. 2018).

5.2 The Abundance Discrepancy Problem

In almost all cases, the brightest emission lines in the optical spectra of photoionised nebulae, like H II regions and PNe, are the various recombination lines of Hydrogen and Helium (predominantly the Hydrogen Balmer series), and lines originating from collisionally-excited transitions of heavier elements like Oxygen, Nitrogen and Sulphur (collisionally-excited lines, CELs). With deeper spectroscopy, much fainter optical recombination lines (ORLs) of these same heavier elements become visible (see, for example, Fig. 5.3). Ultimately, the observed spectra of photoionised nebulae are a complex function of the relative abundances of the various chemical species, and the nebular density and temperature (which are in turn related to the properties of the source of photoionising radiation).

Once corrected for extinction (which can be measured from the ratios of, for example, the Hydrogen Balmer lines), the flux ratios of many observed emission lines can be directly related to the physical properties of the gas—for example, the ratio of [O III] 5007Å to [O III] 4363Å shows a strong dependence on temperature while the ratio of [S II] 6731Å to [S II] 6717Å is very dependent on the electron density (and slightly on the temperature). In empirically deriving the chemical abundances of ionised nebulae, one employs these diagnostic ratios to determine the physical conditions of the gas which can then be used to determine the ionic abundance of a given species (relative to H^+) based on its emission line fluxes (relative to a chosen line of H^+, generally $H\beta$). Total abundances can then be derived either by summing the abundances of the various ionisation states of a given element (where lines of all appropriate ionisation states are available) or by the application of an ionisation correction factor (ICF) which attempts to account for the fractional abundance of a given element in each ionisation state based on the previously derived physical parameters (and often based on detailed photoionisation models; Delgado-Inglada et al. 2014). For a more detailed overview of the entire process of deriving abundances

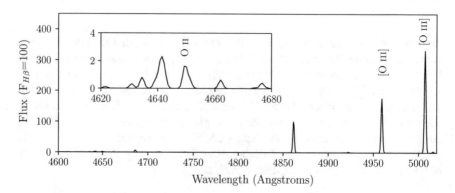

Fig. 5.3 Spectrum of the PN NGC6778, showing the difference in brightness between ORLs (O II 4659 + 4650Å, inset) and CELs ([O III] 4959 + 5007Å) of O^{2+} ion. The scale is such that the flux of $H_\beta = 100$

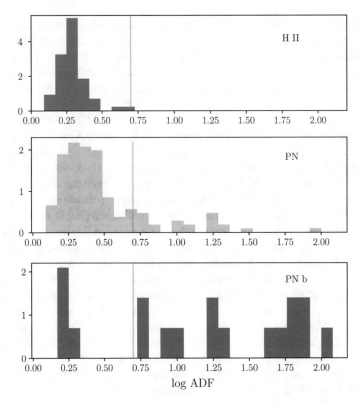

Fig. 5.4 The distribution of the logarithm of the ADF for three different categories of objects: H II (above), PNe (middle), and PNe with known close binary central stars (below). It is clear that almost all extreme ADF systems belong to the latter, hinting at a binary origin. The vertical line corresponds to an ADF = 5, which separates extreme ADFs from normal ones. Based on data from R. Wesson (https://www.nebulousresearch.org/adfs/)

from photoionised spectra, the reader is referred to the excellent review of Peimbert et al. (2017).

The application of the technique outlined above represents a powerful tool to probe astrophysical abundances—frequently used to trace the chemical evolution of stars and galaxies. However, it is not without issue. Beyond the dependence on intrinsically-uncertain ICFs and on the quality of atomic data employed (Juan de Dios and Rodríguez 2017), the derivation assumes homogeneity in temperature, density and chemistry. It has been clear for more than three-quarters of a century that this assumption is flawed (Wyse 1942, a paper we recommend reading, as it shows how with primitive atomic data and measurements from photographic plates, whose non-linear behaviour had to be well understood, one can arrive at surprising results that stand the test of time). In essence, across all photoionised nebulae, abundances calculated using ORLs are systematically larger than those obtained using CELs. This is the notorious *abundance discrepancy problem*—generally characterised by the ratio of ORL to CEL abundances known as the abundance discrepancy factor (ADF[1]). In H II regions, the ADF is on average of order two, which has serious implications for the measurement of the chemical content of galaxies, as these are most often derived using CELs from their ionized interstellar medium (i.e. H II regions). In planetary nebulae, the situation is even more severe, with the majority presenting even larger ADFs and some reaching almost three orders of magnitude larger—the most extreme cases being those of Abell 58 (ADF = 90; Wesson et al. 2008), Abell 46 (ADF = 120–300; Corradi et al. 2015a) and Abell 30 (ADF = 700; Wesson et al. 2003)! Detailed reviews related to this issue can be found in Liu (2006), Peimbert et al. (2017), García-Rojas et al. (2019), and an up-to-date compilation of ADFs can be found on Roger Wesson's web site, https://www.nebulousresearch.org/ adfs/. The distribution of ADFs in these various objects is shown in Fig. 5.4.

We follow here the proposal of García-Rojas and Esteban (2007), see also Wesson et al. 2018) that the origin of the most extreme ADF PNe (i.e. ADF > 5) is different to that of the "normal" ADF PNe and H II regions. Many scenarios have been proposed to explain the extreme ADFs (see the above reviews and references therein), but it generally boils down to the fact the idea that these PNe contain a second gas phase, which is especially bright in ORLs and presents with different metallicity, electron temperature (T_e) and/or density (N_e) to the "standard" PN gas phase, which shines brightest in CELs. It has long been known (Peimbert 1967, 1971; Liu and Danziger 1993) that the electron temperatures derived from the Hydrogen Balmer decrement as well as flux ratios of Hydrogen and Helium I ORLs are systematically below those determined from the CELs.[2] Torres-Peimbert et al. (1980) thus suggested that the large ADFs could be explained by spatial temperature variations. Most recent results indicate that the O II ORLs arise from regions which are approximately a factor of 10 cooler than the regions emitting the [O III] CELs. The likely reason for these cool regions is the existence of a relatively metal-rich (i.e., deficient in Hydrogen) component that allows the gas to cool down extremely efficiently (Liu et al. 2000). In addition to indirect evidence for hydrogen-deficient material found in several PNe, the presence of such material was directly observed in Abell 30 (Wesson et al. 2003) and Abell 58 (Wesson et al. 2008).

Liu et al. (2006) noted that one of the nebulae with the largest known ADF— Hf 2-2—also plays host to a close binary central star with an orbital period of 0.4 days, hinting at a possible connection. Corradi et al. (2015a) later demonstrated that the newly discovered Ou 5, which also hosts a close binary, has a large ADF, as did other PNe hosting close binary stars (Abell 46 and Abell 63). They concluded that *"the most pathological cases"* of strong ADFs *"are planetary nebulae with a close binary central star,"* and predicted *"that many—if not all—of the PNe where large ADFs have been measured are post-CE binaries."* They noted that the reverse was

---

[1]In this chapter, unless otherwise specified, the term ADF refers to the O^{2+} ADF.

[2]The extremely different temperatures derived using using hydrogen ORLs and heavy element CELs clearly demonstrates the flaw in the assumption of temperature homogeneity, as well as the intrinsic issue in deriving all abundances (including froms CELs) relative to the abundance of Hydrogen (in ORLs).

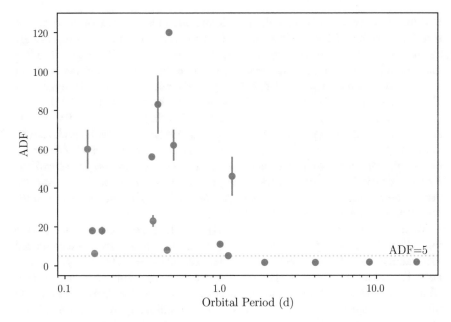

Fig. 5.5 The ADF plotted as a function of the orbital period for all PNe with a close binary CSPN for which the data exist. Based on data from R. Wesson (https://www.nebulousresearch.org/adfs/)

not true, however, i.e. not all systems with a close binary at their centre have large ADF—NGC 5189 and the Necklace being clear examples of "low" ADF PNe with post-CE central stars. This link between strong ADF and close binary central stars was reinforced by Jones et al. (2016) who found that the PN NGC 6778, hosting a 0.15-day binary, showed an average ADF of 18, with the central parts reaching values of 40. This prompted Wesson et al. (2018) to extend the study to a larger sample of nebulae known to host close binary central stars. They detected with FORS2 on the VLT strong recombination line emission in 8 objects (Fg 1, Hen 2-283, Hf 2-2, MPA J1759-3007, NGC 6326, NGC 6337, and Pe 1-9) and all were associated with large ADF values.

With a larger sample of PNe having measured ADFs, Wesson et al. (2018) showed that the distribution of ADFs in PNe could be modeled with two components, one normally distributed and similar to the one operating in H II regions, and the second log-normally distributed and occurring only in a fraction $F = 0.5$ of all nebula, for which the ADF was measured. This second component, corresponds to "extreme ADFs" and may thus be interpreted as due to a second mechanism, operating only in nebulae with close binary central stars. What's more, when looking at the relation between ADF and orbital period, P_{orb} (see Fig. 5.5 for an updated version of their figure), these authors found that there is apparently a threshold orbital period below which the ADF becomes extreme, i.e. all systems with $P_{orb} < 1.15 - 1.2$ d seem to have an ADF > 5, while those with a larger orbital period have ADF compatible with the rest of general PN population. **This appears to be a very strong argument that the extreme ADFs seen in PNe are linked to binarity!** It also argues strongly against the very late thermal pulse (VLTP) scenario as a source of hydrogen-deficient material in PNe, and in favour of nova-like eruptions instead. The limiting period of 1.15 d comes from the fact that the Necklace nebula (not indicated on the figure) has $P_{orb} = 1.16$ d, but no detection of ORLs, which puts an upper limit on its ADF of 5 (Corradi et al. 2015a). On the other hand, Fg 1 has an orbital period of 1.195 d and an ADF $= 46$. The limiting period is thus in between. As stated by Wesson et al. (2018), this finding *"may imply that binaries with periods longer than the $\sim$1.15 d threshold never experience the event which leads to the ejection of Hydrogen-deficient material into the nebula; it could alternatively suggest that in the shorter period objects, the event occurs at the time of or very soon after the ejection of the nebula, while in the longer period objects it may occur later."*

We note, however, that in a system with $P_{orb} = 1.15$ d, the Roche lobe radius (see Eq. 2.1) of the secondary is about 1.5–2 $R_{\odot}$, which is normally too large compared to its radius, so we do not expect mass transfer back on to the WD and thus a possible nova. This, however, ignores the fact that the post-CE companions are often oversized with respect to normal main-sequence stars (see Chap. 3) and in many cases close to filling their Roche lobe (Jones et al. 2015, 2019). Additionally, Wesson et al. (2018) also found an anti-correlation between nebular density and ADF, which would also need to be incorporated in any scenario. This could indicate that the extreme ADFs only occur in those systems that went through the common envelope evolution when the WD progenitor was not too advanced on the AGB, such that the post-CE orbital period is short enough and the nebular electron density not too high (Wesson et al. 2018). The fact that the Necklace, which is known to show Carbon enhancement (see Sect. 5.1)—likely indicating that the progenitor underwent several thermal pulses on the AGB before transferring mass to (and polluting) its companion—has an ADF < 5 supports this hypothesis. In any case, it is useful to note that one could thus use the ADF as a way to discover short-period binaries at the heart of PNe: any object with an extreme ADF should contain one. This is even more interesting, as Wesson et al. (2018) showed that you do not need to detect the ORL lines to establish if a PN has an extreme ADF: they indeed showed that [O II] density diagnostic lines can be strongly enhanced by recombination excitation, while [S II] lines are not, suggesting that one can recognise extreme-ADF PNe as those where $n_e([OII])$ significantly exceeds $n_e([SII])$. This may offer a route towards obtaining a more complete sample of close binary central stars of PNe.

In attempting to understand the origins of extreme ADFs in post-CE PNe, it is interesting to consider the spatial distributions of the ORLs and CELs. As we saw above, extreme ADFs are likely due to the existence of two different gas phases inside the nebula: one of standard temperature and metallicity from which the CELs shine brightest and one cooler, Hydrogen-deficient component which is bright in ORLs. Many studies have shown that the latter seems to be enhanced in the central part of the nebula—primarily by tracing the brightness distributions of ORLs and CELs along the length of the slits used to obtain the nebular spectroscopy (see Fig. 5.6). A finding which has been confirmed two-dimensionally by tunable filter observations,

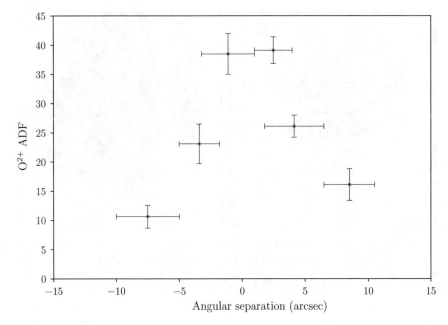

Fig. 5.6 The variation of the O^{2+} ADF along the FORS2 long slit (zero corresponds to the location of the central star) for the PN NGC 6778 (data first presented in Jones et al. 2016)

with OSIRIS on the GTC, of the O^{2+} ORLs and CELs in NGC 6778 demonstrating that the ORL emission appears almost as a nebula-within-a-nebula, originating from a region much closer to the central star than the CEL emitting gas (Fig. 5.7; García-Rojas et al. 2016). This was also observed by García-Rojas et al. (2019) who used MUSE at the VLT to perform integral field spectroscopy and confirm that in several high-ADF PNe, the ORL emission is more centrally concentrated than the emission of the [O III] λ4959 CEL. In some cases, however, slightly different behaviours are seen, such as in Fg 1, where a central peak is also observed, but brighter ORL emission is then also found at the outer edge of the bright central region. As such, although the ORL emitting gas is generally found to be centrally concentrated, consistent with some form of post-CE eruptive event, this hypothesis is clearly not without issue and further investigation is most certainly required.

In conclusion, it seems clear that there must be multiple factors at play in causing the so-called abundance discrepancy problem, with perhaps a combination of both temperature and chemical inhomogeneities responsible for the underlying pattern of ADFs $\gtrsim$ unity found in the general population of PNe and H II regions (García-Rojas et al. 2019). However, it is also similarly clear that an additional ingredient is required to understand the extreme ADFs found in close-binary PNe, which greatly enhances either the temperature or chemical inhomogeneities (perhaps both). Currently, it seems clear that there are observed chemical inhomogeneities in extreme ADF PNe, however the (binary) origins of these are far from certain. Similarly, there appears to

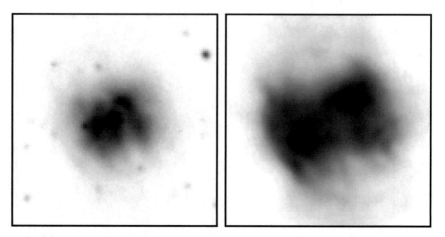

Fig. 5.7 Imagery of the extreme ADF PN NGC 6778 in the light of ORLs (O II 4649+4650Å, left) and CELs ([O III] 5007Å, right), both of which originate from the same chemical species (in this case doubly ionised Oxygen). Note that the ORL emission (left) is more centrally concentrated than the CEL emission (right). Data originally presented in García-Rojas et al. (2016)

be a theoretical footing for the presence of temperature variations (for example, due to the orbital variation of the ionising source inducing resonant temperature fluctuation in the nebula; Bautista and Ahmed 2018) in close-binary PNe, but observational support is still somewhat lacking.

Chapter 6
The Binary Fraction

The principle interest in central star binarity from the PN community is their ability to explain the plethora of observed axisymmetric morphologies (see, e.g., the review of Balick and Frank 2002). However, a critical test of the role of binarity in the shaping of PNe is simply: did enough PN progenitors experience a binary evolution to explain the fact that the vast majority of PNe are not spherical ($\geq 80\%$; Parker et al. 2006)? In this chapter, we will discuss various observational probes of the so-called binary fraction (the percentage of PNe with surviving binary central stars) and the implications for the role of binarity in the formation and shaping of PNe.

6.1 Close Binaries

Given that, as discussed in Chap. 3, the most obvious way to produce a strongly axisymmetric PN via binary interaction is through a common-envelope episode (Chap. 2), then perhaps the most stringent test of the role of binarity in formation of PNe is the close-binary fraction—or the fraction of PNe with surviving post-CE binary central stars.

Already in the 1990s there was an indication that an appreciable fraction of PNe were host to post-CE central stars, through the Bond-Grauer-Ciardullo survey (discussed in more detail in Sect. 3.1.1) which, in spite of its inhomogeneous and biased nature, provided an initial estimate of 10–15% for the post-CE binary fraction. This estimate was significantly refined by Miszalski et al. (2009a) using the OGLE survey—a homogeneous survey of the Galactic bulge—finding a 12–21% post-CE binary fraction based on a sample of 149 PNe covered by the survey. However, this must be considered a lower limit to the true post-CE binary fraction due to the detection constraints of the survey. As highlighted by Jones and Boffin (2017a), systems with lower mass secondaries, longer orbital periods, and/or faint central stars

© The Author(s), under exclusive license to Springer Nature Switzerland AG 2019
H. M. J. Boffin and D. Jones, *The Importance of Binaries in the Formation and Evolution of Planetary Nebulae*, SpringerBriefs in Astronomy,
https://doi.org/10.1007/978-3-030-25059-1_6

could all present low-amplitude variability that would be beyond the detection limits of the survey.

There are indications that the "true" post-CE binary fraction may be much larger than ~20%. The Kepler satellite mission, launched in March 2009 with the aim of photometrically monitoring (with extremely high precision) a region of more than 100 square degrees around the Cygnus-Lyra region, returned useable data on five PNe. Three of these presented variability consistent with a post-CE central star, while a fourth was determined to be the product of a merger event (i.e. having a binary origin which might have had a similar effect on the nebular shaping), indicating a much larger binary fraction of 60–80% (De Marco et al. 2015). Importantly, of the four variables discovered, only one presented with an amplitude that would be easily detectable from the ground, further demonstrating that many systems could have evaded detection in, for example, OGLE. Following the failure of two of the spacecraft's reaction wheels, it was re-tasked to observe multiple fields as part of "campaigns" lasting 80 days each (Howell et al. 2014) with several of the later campaigns covering fields much richer with PNe opening up the possibility of a more robust binary fraction estimation (Jacoby et al. 2018). However, it is important to note that Kepler is far from an ideal instrument to search for photometric variability in CSPNe due to its large pixels (3.98"×3.98") and broad filter band pass (~4300–8900Å) which mean nebular contamination can be rather problematic (see Sect. 3.1.3).

Beyond higher-precision photometry, radial velocity monitoring can also be sensitive to systems that would go undetected by ground-based photometric surveys. While such observations are far more time intensive (to reach a similar signal-to-noise, spectroscopic observations require orders of magnitude more integration time than comparable broadband photometric observations), they have the power to probe much longer orbital periods than photometric monitoring. As summarised in De Marco (2009), several authors have attempted spectroscopic surveys at various resolutions almost always deriving a variable fraction greater than the ~20% from photometry (sometimes as high as 90%). The major caveat being that in most cases the data were insufficient to derive the orbital period and thus rule out other sources of variability, like wind variability (De Marco et al. 2004). More detailed, data-intensive studies have revealed several longer period variables which would elude (and in some cases have eluded) photometric detection (e.g., Manick et al. 2015; Miszalski et al. 2018a, b, 2019a). This hints at a possible missing population of intermediate-period post-CE systems which may prove critical in understanding not only the origins of aspherical PNe but also the common-envelope itself (Sowicka et al. 2017; Brown et al. 2019).

6.2   Long-Period Binaries

Thus far, we have focused on close, post-CE binaries. However, a significant number of binaries with roughly solar-mass, main-sequence primaries will never undergo a CE. For example, roughly one half of the sample of Raghavan et al. (2010) were found to have orbital periods of 100,000 days or longer and thus are unlikely to experience a CE. However, a CE is not necessarily a requirement in order to cause a deviation from single star evolution and, as such, wider binaries may also play an important role in the formation and shaping of PNe (as discussed in Chap. 4).

Due to the extreme difficulty in identifying long-period systems, only a handful are known, and there are almost no observational constraints on the wide-binary fraction. This is principally due to the limitations on the detection of such systems with the majority being discovered either by long-term high-resolution radial velocity monitoring, which is heavily biased towards systems that a bright enough to be studied at reasonable signal-to-noise (e.g., Van Winckel et al. 2014; Jones et al. 2017), or due to their peculiar nature (showing, for example, spectral signatures of a giant rather than post-AGB star; Tyndall et al. 2013; Aller et al. 2018; Löbling et al. 2019, see Chap. 5 for further details).

Some of the very widest central star systems can be resolved using high-spatial resolution imagery. Ciardullo et al. (1999) performed a survey of 113 PNe using the Hubble Space Telescope to search for resolved companions based on close proximity to the central stars, finding ten likely companions (with projected separations less than a few thousand astronomical units) and a further six possible companions (with projected separations of several thousand astronomical units). This results in roughly a 10% wide binary fraction, although the associations are far from certain as are the implied orbital separations (which would in most cases be even wider). In these cases, it is unclear that the evolution of the nebular progenitor would have been in any way affected by the presence of the distant companion, although the analysis of Soker (1999) indicates that, indeed, the presence of a wide companion has in most cases had an appreciable impact on the observed morphologies of the Ciardullo et al. (1999) sample.

6.3 Infrared Excess

One of the *potentially* most promising methodologies for deriving a truly unbiased binary fraction is the search for central star infrared excesses (De Marco et al. 2013). Irrespective of their orbital period (unless so long as to be resolved; Ciardullo et al. 1999), the photometric contribution of a main sequence companion to the spectral energy distribution of the system should always be discernible (except if observed during an eclipse) given high enough photometric precision—i.e. given the clear temperature difference between post-AGB and main-sequence stars, one can discern the presence of lower temperature companions from the excess infrared emission compared to that expected from a post-AGB star alone. As such multi-band, single-epoch photometric observations generally imply a smaller investment of observing time (perhaps as low as five minutes for brighter central stars; De Marco et al. 2013), this opens the possibility to survey a much larger fraction of central stars for these tell-tale signs of binarity. However, the methodology is not without issue. Indeed, while it is, in principle, capable of discovering a significant number of new binary central stars and deriving an unbiased binary fraction, the need for extremely precise photometry means that bright or compact PNe are problematic (and often excluded from the observed samples; Douchin et al. 2015). Furthermore, as highlighted by Barker et al. (2018), the de-reddening process can remove the signal of the companion and cause earlier spectral types to be missed completely. These issues, combined with the intrinsic assumptions on the temperature of the primary and, therefore, its spectral energy distribution, generally mean that the methodology can only be claimed to select binary *candidates* rather than to detect binarity (Barker et al. 2018).

In spite of the difficulties of searching for central star infrared excesses, the technique has been successfully applied in order to derive a binary fraction with the initial work of De Marco et al. (2013) finding a value of 67–78% based on I-band excess and a fraction of 100-107% based on excesses detected in more sensitive J-band observations (where lower temperature companions would become clearer). This was then built upon by Douchin et al. (2015), employing a refined measurement technique and increased sample, to derive a fraction of $40 \pm 20\%$ for the I-band and $62 \pm 30\%$ for the J-band. Furthermore, they noted that white dwarf companions would not be included in this measurement, meaning that likely an additional 13–21% should be added to the measured fractions in order to derive the true binary fraction. While the uncertainties are large, these values are certainly indicative of a far greater binary fraction that the $\sim$20% found by photometric variability monitoring (Miszalski et al. 2009a), strongly supporting the hypothesis that a significant population of binary central stars (at longer orbital periods, or with lower mass companions) has, thus far, evaded detection (Jones and Boffin 2017a). Importantly, while Douchin et al. (2015) found that SDSS data was not of sufficient quality to adequately account for nebular contamination in the z-band, the initial results using VPHAS+ (Drew et al. 2014) were more promising, detecting i-band excesses in 3 of the 7 central stars in the final sample of Barker et al. (2018). This opens the door for a future estimation of the binary fraction via infrared excess using a much larger sample of central stars covered by all-sky surveys, like that of the Large Synoptic Survey Telescope (LSST; Ivezić et al. 2008).

6.4 Planets, Mergers and Higher Order Systems

While all of the previous discussion has focused on binarity, it is important to note that more than 10% of solar-type stars are found in triples or higher order systems (Raghavan et al. 2010) and, as such, these systems may also play an important role in the formation and evolution of PNe. To date, only one confirmed triple central star has been discovered—that of NGC 246 which was found by Adam and Mugrauer (2014) to be a wide, co-moving triple. However, Bear and Soker (2017) hypothesise that interacting triples may be responsible for the formation of morphologies which show a marked deviation from axial and/or mirror symmetry, and following by-eye classification of observed PN morphologies, find that some 13–20% of all PNe may result from triple interactions. Here it is important to note that in many scenarios the triple will not survive the formation of a PN, with one or more stars being ejected from the system or merging (Soker 2016; Jones, Pejcha & Corradi 2019).

Mergers may similarly play an important role in the formation of PNe from binary systems, or from stars with planetary mass companions which are particularly unlikely to survive a common-envelope phase (given the limited orbital energy available to unbind the envelope). In spite of not surviving the event, the companion (planetary or stellar mass) will deposit angular momentum into the envelope of the PN progenitor (Soker 1997) and possibly even lead to the formation of jets (Soker 1996), impacting on the morphology of the resulting PN (albeit perhaps only slightly; De Marco and Soker 2011; Staff et al. 2016). To date, no surviving planetary system has been found in orbit around the central star of a PN, consistent with the picture that most planets will either be destroyed or have their orbits expanded to regimes that make their detection challenging (Mustill and Villaver 2012), particularly given the intrinsic difficulties in studying the variability of PN central stars (see above and Chap. 3). Several planetary candidates have been found in orbit around post-CE binaries, but it is unclear whether these survived the CE or are second-generation planets formed from the ejected envelope (Guzman-Ramirez et al. 2014; Hardy et al. 2016).

Only one PN central star has been identified as the product of a merger—that of NGC 6826 found by De Marco et al. (2015) to be rotating too fast to have a single star origin. However, while merger remnants should always be bipolar in nature (MacLeod et al. 2018), it is not clear that all will result in the formation of a planetary nebula—depending on the masses and evolutionary phases of the two binary components, one might expect a luminous red nova (Metzger and Pejcha 2017; Blagorodnova et al. 2017), a supernova (Kashi and Soker 2011; Maoz et al. 2014) or other exotic objects like R Coronae Borealis stars (Clayton et al. 2011).

While all of the aforementioned scenarios may be important in explaining the observed morphologies of PNe, there are essentially no observational constraints on the fraction of PNe derived from planetary interactions, stellar mergers or triple star systems.

6.5 Theoretical Expectations

Given the wide-range of observational binary fraction estimates, ranging from $\sim 20\%$ for the photometrically-detectable close-binary fraction through to 60–80% based on radial velocity variability and infrared excesses, it is interesting to contrast these values with theoretical expectations. Repeating the somewhat back-of-the-envelope calculations of Boffin (2015), while the binary fraction of all solar-like stars is roughly 50% (Raghavan et al. 2010), the fraction with companions close enough to experience some kind of interaction (i.e. separation ≤ 200 au) drops to roughly 30%. One may restrict this further, to only the systems that would be likely to experience a CE which, based on a maximum AGB radius of 1000 $R_\odot$, Boffin (2015) restricted their estimate to systems with orbital periods less than 21–30 years. The studies of Duquennoy and Mayor (1991) and Raghavan et al. (2010) find the orbital period distribution of solar-type stars to be roughly gaussian, allowing one to derive the number of systems with orbital periods less than the ~30 year cut-off to be approximately 30% (consistent with the observed fraction of white-dwarf-main-sequence binaries discovered by SDSS that were found to be post-CE systems, at ~35%; Schreiber et al. 2008). As the total binary fraction is ~50%, this would imply a post-CE fraction of ~15% assuming no systems merge (but see also the discussion in Chap. 4). This is seemingly consistent with the 12–21% fraction derived observationally by Miszalski et al. (2009a), but as highlighted earlier a significant number of systems could have evaded detection and, furthermore, it is highly unlikely that *no* systems merge during the CE.

The population synthesis work of Moe and De Marco (2006) highlighted that if all low- and intermediate-mass stars experienced a PN phase the total number of observable PNe in the Galaxy would be significantly greater than that observed. They go on to argue that this issue can be reconciled if only a fraction, roughly consistent with the binary fraction on the main-sequence, went on to produce PNe—i.e., **that only binaries go on to produce PNe**. This idea has been claimed to be consistent with observations of globular clusters where binary interactions are often invoked as the only way a PN could form in such old stellar populations (Jacoby et al. 2013; Bond 2015). However, recent models of post-AGB evolution indicate that even the very low-mass remnants that would be found in these regions will evolve fast enough after leaving the AGB to produce an observable PN (Miller Bertolami 2016), seemingly negating the need for a binary pathway toward PN formation in globular clusters (but, see Chap. 8). Other population synthesis attempts place the binary fraction at a more conservative value (e.g. ~40%; Han et al. 1995), however given the number of assumptions that are required in order to perform such population syntheses as well as the large uncertainties surrounding various critical phases of stellar evolution (particularly the common envelope prescription and efficiency employed; Toonen and Nelemans 2013, see also Chap. 2) it is unclear how much one can rely on these estimates (Toonen et al. 2014).

6.6   Conclusions

While both observational and theoretical estimates of the binary fraction among the central stars of PNe vary wildly, it is clear that the fraction is appreciable (at least 20%). Furthermore, in understanding the origins of PNe, the observed binary fraction itself does not come close to providing the complete picture. Some binary companions, though still present, may have been in too wide orbits to have any noticeable effect on the evolution of the nebular progenitor, while others (including planetary mass companions) may have played a much greater role via merger and would now no longer be counted in the binary fraction.

Chapter 7
Post-AGBs and Pre-planetary Nebulae

When a star ascending the asymptotic giant branch (AGB) has lost most of its envelope (via stellar wind or mass transfer), it will progressively evolve to the blue, as the core becomes more and more exposed. At a given moment, the central star is hot enough ($T \sim 30,000\,\mathrm{K}$) to ionise the surrounding envelope and a planetary nebula (PN) forms. There is thus a fleeting moment, that lasts about 10% of the PN phase itself, where the star has left the AGB but there isn't a PN yet—the nebula that surrounds the star isn't yet ionised, and can only be seen in scattered light. When visible, the central star is generally characterised as a supergiant of spectral type F or G. This evolution at constant luminosity, but increasing effective temperature (see Fig. 1.5), corresponds to the post-AGB or pre-planetary nebula (PPN[1]) phase. Given the relatively short duration of this phase (around 1 000 years), it is clear that PPNe are relatively rare.

Historical reviews on PPNe and post-AGBs can be found in Kwok (1993) and van Winckel (2003), respectively, while a recent and up-to-date review on post-AGB is available in Van Winckel (2018). The distinction between PPNe and post-AGBs is hard to make. Van Winckel (2018) makes a distinction based on whether the objects show a resolved reflection nebula or not. If they do they are called PPNe and these will definitively evolve into planetary nebulae (Sahai et al. 2007). Post-AGB star is an umbrella term which includes PPNe but also many objects that will not ionise the surrounding matter (i.e. not form a PN) as it is too dispersed by the time the central star becomes hot enough (see also Lagadec 2018). These latter objects will instead evolve into, e.g., Barium stars (see Chap. 4). Another way to look at this is whether the post-AGB has a disc or a detached shell (van Winckel 2003). The latter—PPNe—can

---

[1]Historically, the name "proto-planetary nebula" was more commonly used (e.g., Kwok 1993) however, as highlighted by Sahai et al. (2005), the term "proto-planetary" is now widely used to refer to discs around pre-main sequence stars. As such, the continued use of proto-planetary nebula unfortunately has ambiguous implications (especially when one also folds in the inherently confusing nature of the misnomer "planetary nebula" to begin with!), and it is therefore important to use the (slightly) more favourable name "pre-planetary nebula".

© The Author(s), under exclusive license to Springer Nature Switzerland AG 2019
H. M. J. Boffin and D. Jones, *The Importance of Binaries in the Formation
and Evolution of Planetary Nebulae*, SpringerBriefs in Astronomy,
https://doi.org/10.1007/978-3-030-25059-1_7

be distinguished easily from the former by their spectral energy distribution (SED), especially in the infrared: detached shells present an SED without a near-IR excess and a peak in the SED typically around 30–60 μm (Van Winckel 2018).

7.1 Post-AGBs, Part II

We consider here those post-AGBs that are thought (or shown) to have a circumbinary disc, with little or no extended nebulosity—also called disc post-AGBs (or dpAGB, when space is an issue). From a survey in the Magellanic Clouds, Kamath et al. (2014, 2015) infer that more than half of optically bright post-AGBs are those with a clear near-infrared excess. Such an excess is established to be linked to the presence of a stable, Keplerian disc, which have been directly observed in a few objects using adaptive optics, interferometry and/or with ALMA (Bujarrabal et al. 2015; Hillen et al. 2016; Bujarrabal et al. 2016, 2017; Ertel et al. 2019). The ALMA observations of IW Car by Bujarrabal et al. (2017) reveal the presence of an equatorial disc, as well as slowly outflowing gas in an hourglass-like structure whose symmetry axis is perpendicular to the rotation plane and is probably comprised of material extracted from the disc. Another famous example of such an object[2] is the Red Rectangle (Bujarrabal et al. 2013, 2016, see Fig. 7.1), which is a reflection nebula with a bipolar shape. It is known to harbour both a disc from which jets are launched, as well as an interacting binary with an orbital period of 319 days and an eccentricity of 0.38. The primary feeds an accretion disc around the secondary, from which the bipolar jets originate (Van Winckel 2014). In several of these post-AGB stars, indirect evidence has been found for high velocity, weakly-collimated outflows or jets that originate from around the main sequence companion (Gorlova et al. 2012; Bollen et al. 2017).

Quite interestingly for our purpose, radial velocity surveys have shown that **basically all objects in this category are binaries**, with orbital periods between 100 and several thousand days (Van Winckel et al. 2006; Van Winckel 2007; Oomen et al. 2018). These orbits are too small to accommodate the large radii of AGB stars, meaning that these systems underwent some shrinking of their orbit, but still managed to avoid a dynamic event that would have decreased their orbital periods to a few days or a few hours (see Chap. 4). So either the mass transfer was stable, or if a common envelope formed, it was very quickly expelled. The presence of a large Keplerian disc whose total angular momentum is not negligible could, provided that all momentum comes from the binary system, lead to a significant decrease of the distance between the stars (Bujarrabal et al. 2016). The disc may also explain why these systems are not circular (Dermine et al. 2013). Bujarrabal et al. (2018) found that IRAS 08544-4431 has very similar properties to the Red Rectangle. By comparing the current angular momentum in the disc to that of the binary system, these

---

[2]One should note, however, that the Red Rectangle is atypical is having a nebula surrounding it!

Fig. 7.1 The Red Rectangle nebula is a remarkable, and perhaps atypical, example of a disc post-AGB object. The unusual shape of the Red Rectangle is likely due to a thick dust torus which pinches the outflow into tip-touching cone shapes. Credit: ESA/Hubble & NASA

authors conclude that the orbit must have shrunk by a factor $\gtrsim 2$. Alternatively, the jets present in these systems may play a role in avoiding the common-envelope phase (Soker 2017).

7.2 Proto-Planetary Nebulae

Observed PPNe appear to have rather remarkable shapes, with a large fraction presenting either bipolar or multipolar morphologies, and almost all the rest showing some form of elongation or deviation from sphericity (Sahai et al. 2007; Lagadec et al. 2011). AFGL 2688 (Fig.7.2), the first PPN ever discovered, presents a set of concentric rings around the central star, as well as a very thick layer of dusty gas and bipolar outflows. It is this circumstellar material that imposes very large self-extinction, offering a clear explanation as to why such objects were only discovered relatively recently, thanks to infrared observations—initially by *Air Force Geophysical Laboratory (AFGL)* rockets. In the case of AFGL 2688, the onion-like diffuse structure surrounding the object appears to have been produced on timescales of a few hundred years, much too short to be linked with thermal pulses that occur on timescales of 10^4 years at best. This is, however, very reminiscent of the spiral

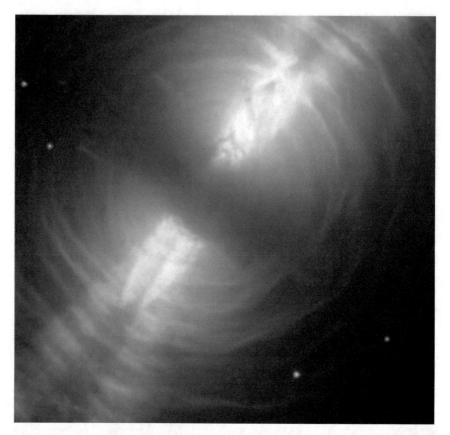

Fig. 7.2   The PPN AFGL 2688, also known as the *Egg Nebula*, as observed with the WFC3 attached to the Hubble Space Telescope. Credit: ESA/Hubble & NASA

structure seen around the AGB star R Sculptoris (Maercker et al. 2012) and whose spacing of 350 yrs is thought to be due to a binary system with such an orbital period. The companion could also be a brown dwarf in some cases, as inferred for the origin of the spiral around the AGB star EP Aquarii (Homan et al. 2018).

These PPNe generally harbour a torus, relatively massive and slowly expanding, which may be the origin of the often observed jets. Thus, the properties of PPNe seem to imply that the agents responsible for the unusual shapes of PNe are already in place at the end of the AGB phase, or at least at the beginning of the PPN phase (Sahai and Trauger 1998). Indeed, Sahai and Trauger (1998) proposed a model where fast jets are triggered at the end of the AGB phase and give the PN a bipolar shape. They posit that the jets could arise through mass transfer from a companion. This hypothesis clearly supports the idea that these PPNe are the progenitors of aspherical PNe. However, it is important to highlight here that the observed morphologies of PPNe are not necessarily reflective of the actual physical structure with illumination effects playing an important role in relating the two (Koning et al. 2013). Furthermore, the

Fig. 7.3 Three examples of proto-planetary nebulae: the Westbrook Nebula (AFGL 618; top left), IRAS 13208-6020 (top right), and the Calabash Nebula (OH 231.8 + 04.2, bottom). Credit: ESA/Hubble & NASA

Fig. 7.4 The young PN M2-9, also known as "Minkowski's Butterfly Nebula" or the "Twin Jet Nebula". Credit: ESA/Hubble & NASA; Acknowledgement: Judy Schmidt

hydrodynamic models of Huarte-Espinosa et al. (2012) demonstrate that an initially bipolar (or even multipolar) PPNe can evolve towards more elliptical shapes as they mature as PNe—meaning that there is no obvious or direct correlation between observed PPN and PN morphologies.

As we have repeatedly seen, torii and jets are typical consequences of binary interaction. Indeed, Sánchez Contreras et al. (2004) found that the central star of the Calabash nebula (Fig. 7.3) has a companion, likely a main-sequence star with spectral type A. On the other hand, despite a long-term monitoring of seven bright PPNe spanning 25 years, Hrivnak et al. (2017) couldn't detect any clear orbital motion, with the possible exception of one candidate. They conclude that if these objects are binaries, they must be of low mass, $\leq 0.2\,M_{\odot}$, or long period, >30 years. The objects they studied are, however, undergoing pulsations (with periods of 35–135 days) that *induce complex atmospheric differential motions*, whose signatures in the photometry and radial velocities are clear, making the detection of a wide binary very challenging. Moreover, the residuals of their radial velocities when the pulsation signal is removed are rather large compared to the individual measurement errors. This could be due to the orbital motion of a long period binary.

As mentioned above, the bipolar and multipolar shapes of PPNe are likely the product of binary interaction. Perhaps, these are thus very wide systems, as seen for example in the young planetary nebula Minkowski's Butterfly Nebula (M 2-9; Fig. 7.4). This amazing object presents two thin outflows with a narrow waist, and the presence of a dense torus. Repeated observations over several years (Corradi et al. 2011a) showed a lighthouse effect, that is, a pattern that precesses around the symmetry axis of the nebula, indicating an interacting symbiotic-like binary system at the centre of the nebula with an orbital period of 86 ± 5 years. If such binaries are at the centre of PPNe, current observations wouldn't be able to detect them. Given the orbital period distribution of solar-like binaries (see Chap. 4), which peaks at about 350 yrs, such periods are not unexpected. What kind of mass transfer could then explain these features is, however, still a matter of debate (Blackman and Lucchini 2014).

While it is feasible that many of these PPNe host long period binaries, the absence of close binaries does seem somewhat troubling given that the general population of PN central stars is known to comprise at least 20% close binaries (see Chaps. 3 and 6). Here, it is perhaps important to highlight that post-CE central stars might feasibly experience an accelerated evolution with respect to isolated stars, as the stars coming out of a CE are far from thermal equilibrium and their Hydrogen envelopes may be smaller than those of post-AGB stars computed using standard stellar wind prescriptions (Miller Bertolami 2017). **Thus, we should perhaps not expect close, post-CE binaries at the centre of PPNe.** Ultimately, our poor understanding of the CE phase limits our ability to reach reasonable conclusions regarding the apparent absence of post-CE central stars in PPNe.

Chapter 8
Binarity and the Planetary Nebula Luminosity Function

As described in Chap. 1, the spectra of planetary nebulae (PNe) are characterised by bright emission lines, from Hydrogen and Helium, but also from other elements (Fig. 8.1). In particular, the forbidden lines of Oxygen are very strong, and this is particularly true for the [O III] $\lambda5007$ line (the 'nebulium' of the early days). According to Dopita et al. (1992), at high metallicities, and for the hottest central stars, up to 15% of the luminosity of the central star is re-emitted in this line alone[1]! As such stars are rather luminous (several hundreds to thousands of times the luminosity of the Sun), this means that when seen through a narrow-band filter that isolates this line, PNe can be seen from very far away, without the need to resort to spectroscopy.

Baade (1955) was the first to observe PNe in another galaxy, Andromeda (M31), while Ford (1978) performed deeper observations in the same galaxy and detected hundreds of PNe. A more recent census of Andromeda was done by Bhattacharya et al. (2019) who identified no fewer than 4289 PNe! Studying PNe in other galaxies offer the advantages to be able to make a complete census of the population, unlike in the Milky Way where we are limited to study outside the dusty plane. Moreover, PNe are linked to low- and intermediate-mass stars and are therefore good tracers of stellar populations (Renzini and Buzzoni 1986). PNe can be used as tracers of light, chemistry and kinematics in galaxies (Meatheringham et al. 1988; Hui et al. 1995; Arnaboldi et al. 1996; Longobardi et al. 2018).

Hodge (1966) was most likely the first to suggest that planetary nebulae could be used as standard candles to determine extragalactic distances. Later, Ford et al. (1978) used PNe to assert that M32 must be in front of M31, while Ford and Jenner (1978) used PNe to get the relative distance between M31 and M81. Observations in more galaxies revealed that the brightest PNe in these had similar absolute fluxes

---

[1]This decreases to 5% for low metallicity stars and even less for the coolest central stars. Some more recent works seem to indicate that the maximum amount of the reprocessed light is 12%.

© The Author(s), under exclusive license to Springer Nature Switzerland AG 2019
H. M. J. Boffin and D. Jones, *The Importance of Binaries in the Formation and Evolution of Planetary Nebulae*, SpringerBriefs in Astronomy,
https://doi.org/10.1007/978-3-030-25059-1_8

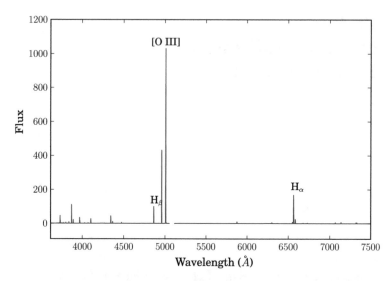

Fig. 8.1 FORS2 spectrum of a planetary nebula, covering a wide spectral region. The scale is such that the flux of $H_\beta = 100$. Note the very strong lines of [O III]

in the [O III]λ5007 line, seemingly indicating that there was a limiting maximum brightness amongst PNe and confirming their possible role as standard candles.

It was soon discovered (Ciardullo et al. 1989; Jacoby 1989) that such studies could be made more rigorous by looking at the planetary nebula luminosity function (PNLF), that is, the relative number of PNe when divided into [O III] magnitude bins. The evolution timescale and the intrinsic luminosity of a PN will depend on the mass of the progenitor, thus the PNLF is the result of the mass distribution as well as the stellar and nebular evolution (Kwok et al. 2000).

The [O III] λ5007 magnitude (M_{5007}) of a PN is related to its monochromatic flux by

$$M_{5007} = 2.5 \log F_{5007} - 13.74,$$

where F_{5007} is the absolute flux of the nebula in the light of [O III] λ5007 when viewed from a distance of 10 pc, expressed in ergs cm^2 s^{-1} (Jacoby 1989).

Although the faint end can vary quite a lot from one galaxy to another (see, e.g., Bhattacharya et al. 2019, for a recent example), the bright cut-off of the PNLF appears to be an invariant (see Fig. 8.2). The PNLF was modelled by Ciardullo et al. (1989), based on a first estimate by Henize and Westerlund (1963), as

$$N(M) \propto e^{0.307\, M} \left(1 - e^{3(M^* - M)}\right),$$

where M^* is the magnitude of the bright cut-off (the limiting magnitude brighter than which no PNe are observed). The value of M^* was initially estimated from one hundred PNe in Andromeda. A revised estimate gives $M^* = -4.54 \pm 0.05$ (Ciardullo

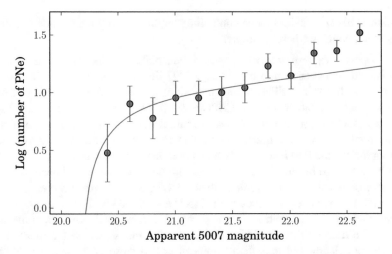

Fig. 8.2 The PN luminosity function of M31, based on the data of Ciardullo et al. (1989) for their homogeneous sample. The apparent cutoff used in the modelled curve used here is $m^* = 20.17$. For a modern version of the PNLF in M31, see Bhattacharya et al. (2019)

2013), with indications of a slight dependence on M* on metallicity (Dopita et al. 1992; Schönberner et al. 2010). Jacoby (1989) found that such a bright cut-off could be explained if one assumed a very narrow distribution of central star masses, i.e. a Gaussian distribution centred on $M = 0.61\ M_\odot$ and with a sigma, $\sigma = 0.02\ M_\odot$!

Longobardi et al. (2013) generalised the functional form of the PNLF, introducing two parameters, c_1 and c_2, where the first one is simply a normalisation constant, while the second is apparently correlated with the star formation history of the parent stellar population:

$$N(M) = c_1 e^{c_2\,M} \left(1 - e^{3(M^*-M)}\right).$$

Bhattacharya et al. (2019) found for M31 that $c_2 = 0.279 \pm 0.024$, i.e. in agreement with the value used by Ciardullo et al. (1989). It is thus remarkable that after a tenfold increase in the number of PNe detected in Andromeda, the initial result obtained 30 years ago is still valid!

Quite amazingly, the shape of the bright end of the PNLF appears to follow this relation in a variety of stellar populations, both in the discs and bulges of spiral galaxies, but also in elliptical and irregular galaxies. To date, this method was used to determine the distances to more than 60 galaxies, up to and including inside the Virgo and Fornax clusters, and is key part of the extragalactic distance ladder. **The reason why this is amazing is that there is no real theoretical understanding of why this would work at all.** This is reminiscent of another well-known standard candle, Type Ia supernovae, used with increasing accuracy to determine the Hubble constant, even though it is still unclear how these supernovae are formed! In fact,

Ciardullo (2012)[2] provides some compelling arguments why the method shouldn't work at all, which we hereby resume:

- The observed value of M^* corresponds to about $600\,L_\odot$, which, assuming a realistic stellar flux conversion into the [O III] $\lambda5007$ line, implies that the brightest PN have a central star (CSPN) with luminosities above $\sim5,500\,L_\odot$, that can only be produced by rather massive CSPNe of $\sim0.6\,M_\odot$. These, on the other hand, require massive progenitor masses, i.e. initial masses $>1.9\,M_\odot$. As such stars have relatively short lifetimes, we do not expect them in older stellar populations (even though the PNLF seemingly holds even in these regimes).
- To have enough bright PNe as observed, elliptical galaxies would need to have a sizable population of stars younger than ~1 Gyr, which is not observed.
- There is also reason to expect that the bright end of the PNLF should evolve with time, and not be the invariant that is observed. Indeed, as the masses of the stars that have finished their evolution changes with time, so would the mass of the CSPN population and thus their luminosities and those of their host PNe. The expectation is that over a timescale of 10 Gyr, the decline in brightness should be more than 4 mag!
- Finally, most bright PNe in the Milky Way (Fig. 8.3) are asymmetric, some presenting with dust lanes and highly variable self-extinction that should affect their measured brightnesses. This apparent non-uniformity is difficult to reconcile with being a distance indicator.

Yet, the PNLF appears to be able to provide distances to galaxies at the 10% level! Ciardullo (2012) thinks that, here again, the solution is to be found in binary evolution, and that mergers or mass transfer could produce stars massive enough to account for the observed bright PNe, even in old systems. Whether this is indeed the solution will most-likely necessitate the identification of the brightest PNe in the Milky Way, that is, being able to measure precisely their distances. Hopefully, the Gaia DR3 will provide the required information. Detailed study of these brightest PNe can then be used to test the hypothesis that they are the products of binary evolution.

Very recently, Gesicki et al. (2018) provided a different solution. They assert that the existence of an invariant M^* could be simply explained thanks to new evolutionary tracks of low-mass stars. They produced simple photoionisation models based on a model grid of seven of the new evolutionary tracks of Miller Bertolami (2016)—covering initial masses in the range 1–3 $M_\odot$ at a metallicity $Z = 0.01$. All their models assumed that the PNe were comprised of spherical shells of total mass $0.1\,M_\odot$, in which the shell expands changing from optically thick to optically thin. By varying the assumed kinematics they derive two scenarios: the intermediate-nebula hypothesis (whereby the nebulae remain optically thick for the majority of their evolution) and the minimum-nebula hypothesis (where the PNe are predominantly transparent). The [O III] $\lambda5007$ flux of each model was then measured for each timestep of their model grid. The final modelled PNLFs were then derived by integrating over the stellar mass distribution (assuming a Salpeter initial mass function) and

---

[2]See also references therein, as well as Davis et al. (2018).

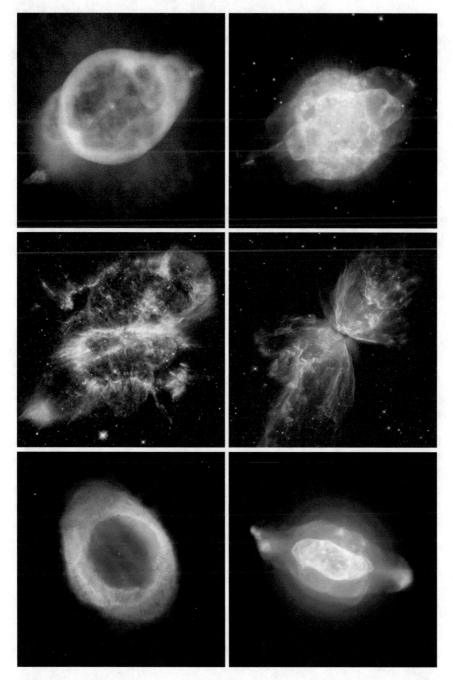

Fig. 8.3 Some examples of intrinsically bright planetary nebulae in the Milky Way, showing that they are rather asymmetric and may have a non-negligible self-extinction. From top to bottom and from left to right: NGC 3242, NGC 3918, NGC 5189, NGC 6302, NGC 6720 and NGC 7009. Credit images: NASA, ESA and ESO

interpolating between the stellar models. Their integrated PNLFs all present with a strong cut-off which is relatively close to the observed value of M* as long as the PNe remain optically thick for the majority of their evolution (the minimum-nebula case produces cut-offs which quickly move to fainter values with the age of the population). Clearly, the conclusion that Gesicki et al. (2018) reach based on these models—principally that the PNLF can be reproduced by single stars—still has a few important sticking points:

- All their model PNe were spherical shells of the same mass with expansion rates which mean they remain optically thick for a significant fraction of their evolution. This is clearly not valid for an entire population of PNe, but may serve at least as an indication of validity.
- In their models, the bright cut-off is always the result of progenitors with initial mass in the range 1.1–2 $M_\odot$—which is still rather massive for the very oldest populations in which the PNLF has still be shown to be invariant.
- Davis et al. (2018) studied the (self-)extinction that affects PNe in M31's bulge as well as in some other galaxies, and showed that this was clearly non-negligible, leading to a higher value of M^* (by up to 1 mag)! There is no way to produce such bright PNe, even using the new evolutionary tracks, with progenitors less massive than 2.5 $M_\odot$, which are relatively short-lived. There are not enough of these stars in M31's bulge to be able to explain the number of observed bright PNe.

Binary evolution may offer a route towards resolving these issues, where mergers and/or mass transfer could produce more massive progenitors even in old populations. Furthermore, as highlighted by Gesicki et al. (2018), the departure from sphericity generally associated with central star binarity could help maintain the PNe opaque for longer during their evolution. Previous three-dimensional modelling by Gesicki et al. (2016), with PNe comprising dense equatorial torii (which would remain opaque for longer) and thin lobes are found to reproduce well observed line emissivities including those of [O III]. Ultimately, given that such a significant fraction of PNe must be the result of a binary evolution, it is perhaps not surprising that binarity might be the key to understanding the PNLF. However, much work is left to be done before the exact role of binary stars in the PNLF is truly understood.

Chapter 9
Conclusions and Outlook

In attempting to draw conclusions with regards to the overall importance of binarity in the formation and evolution of PNe, it is perhaps appropriate that we return to the syllogisms used by Bond (2000) to demonstrate the need for caution:

Planetary Nebulae

1. Binaries eject axisymmetrical PNe.
2. Most PNe are axisymmetrical.
3. Therefore most PNe are ejected from binaries.

Ducks

1. Ducks quack.
2. I quack.
3. Therefore I am a duck.

The syllogism on the left does indeed seem plausible, but that does not mean that it is logically sound as made clear by considering the equivalent syllogism on the right. However, one can make the case that in the last twenty or so years, since this argument was initially raised by Bond (2000), the weight of (albeit still generally circumstantial) evidence in favour of the statement "Most PNe are ejected from binaries" has grown significantly. This is particularly clear when one considers that in the year 2000 there was/were:

- Still no robust estimate of the surviving binary fraction (neither close nor total; Miszalski et al. 2009a; Douchin et al. 2015).
- Only thirteen binary central stars with confirmed periods[1] (Bond 2000).
- Still no known binary central star with an orbital period longer than a few days[2] (Van Winckel et al. 2014; Jones et al. 2017).

---

[1] In fact, this number should really be only twelve given that one of the thirteen listed by Bond (2000) has since been shown to most-likely be a chance alignment with a binary field star (Jones and Boffin 2017b).

[2] Although a couple of central stars had been found to display composite spectra which implied longer orbital periods (due to the evolved nature of the cooler components; Bond and Livio 1990).

© The Author(s), under exclusive license to Springer Nature Switzerland AG 2019
H. M. J. Boffin and D. Jones, *The Importance of Binaries in the Formation and Evolution of Planetary Nebulae*, SpringerBriefs in Astronomy,
https://doi.org/10.1007/978-3-030-25059-1_9

- still no clear indication of the ubiquity of hot and/or warm Jupiters[3] (large planets on short-period orbits which could feasibly impact the evolution of their host stars; Huang et al. 2016).

9.1 Just How Important Is Binarity?

Trying to exercise appropriate caution, let us then attempt to qualify the importance of binarity in the formation and evolution of PNe.

9.1.1 Can Single Stars form Planetary Nebulae?

The most obvious place to start is by considering the most extreme version of the so-called "binary hypothesis". Is it possible that central star binarity is a necessity for the formation of a visible PN? This may seem like an outlandish statement but it may not be so ridiculous given that, for example, binary fraction estimates based on J-band infrared excesses are very nearly consistent with 100% ($62 \pm 30\%$; Douchin et al. 2015). Furthermore, the population synthesis efforts of Moe and De Marco (2006) indicate that the number of observed Galactic PNe is discrepant with the number expected by almost 3σ—with those authors concluding that only a subset of stars generally thought capable of producing a PN actually do, hinting that the population might be better explained if *only* binaries produce PNe. However, the visibility of a PN is a function of its own kinematics and density (i.e., the time it takes for the shell to disperse into the surrounding ISM), as well as the temperature and luminosity of its central star (i.e. is the star hot and luminous enough to ionise the nebula?). Previously, post-AGB evolutionary models indicated that almost irrespective of the nebular properties, low-mass stars would take far too long to reach $\log T_{eff} \geq 4.4$ and would thus never produce a visible PN (Schoenberner 1983; Vassiliadis and Wood 1994). Modern calculations incorporating improved microphysics alter this interpretation—instead indicating that even rather low-mass stars reach such high temperatures on short enough timescales (Miller Bertolami 2016). This does not necessarily mean that even more PNe should be observed, as the more rapid post-AGB evolution also means that higher mass stars cool below the required temperature more quickly. This situation becomes even more complicated when one considers that depending on the kinematics, the shell may be opaque or already transparent to the ionising radiation at that point in its evolution. In any case, it seems exceptionally unlikely that *no* single star can satisfy the required criteria of having a high enough temperature and luminosity while its ejecta is still opaque—therefore it seems safe to assume that at least some PNe are formed from single stars.

---

[3]The stellar binary fraction, however, was already relatively well-constrained for solar-type stars (Abt and Levy 1976).

9.1.2   Can Single Stars form Bipolar Planetary Nebulae?

Now that we have established that at least some single stars must form PNe, we have to ask whether single stars are capable of reproducing those morphologies classically associated with binarity. Taking the extreme case of forming an hourglass—similar to that of the archetypal PN MyCn 18 (see Fig. 9.1)—there is a requirement that the shaping agent (or combination of shaping agents) must be able to embed sufficient axisymmetry in the ejecta as described in García-Segura et al. (2014). The two clearest (non-binary) contenders for imparting such axisymmetry are magnetic fields and/or rotation. Nordhaus et al. (2007) studied in detail the evolution of magnetic fields throughout the AGB in both an isolated (single star) setting and a setting in which a low-mass companion is embedded inside the envelope. Their simulations show that in isolated stars, the magnetic dynamo is unsustainable as turbulent dissipation and Poynting flux act to drain the required rotation and differential rotation. A similar conclusion was reached by García-Segura et al. (2014), who demonstrated that for single stars even if the magnetic torques are only turned on at the end of the star's

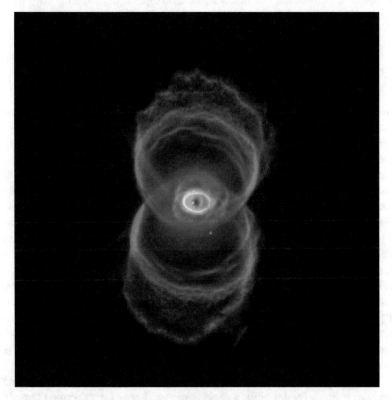

Fig. 9.1 HST image of the Etched Hourglass Nebula—MyCn 18. Image credit: Raghvendra Sahai and John Trauger (JPL), the WFPC2 science team, and NASA/ESA

evolution they are still incapable of carrying enough angular momentum to the stellar surface to reproduce hourglass-like morphologies. They further demonstrated that rotation alone (without any magnetic-field-induced angular momentum transport) is generally expected to be several orders of magnitude too weak to impact on the mass-loss morphology.

One may, in light of the vast numbers of exoplanets discovered, expand the definition of single star to a, perhaps, more realistic case of a star with no stellar companion but one or more sub-stellar companions in relatively close orbits. In this case, the sub-stellar companions can be a source of angular momentum but, as highlighted by Staff et al. (2016), are unlikely to provide enough to reseed the differential rotation required to drive the dynamos considered by Nordhaus et al. (2007). If the planet is engulfed and tidally-disrupted before merging with the core, it is possible that a disc-driven outflow may form (Soker 1996). This may help to drive axisymmetry but would almost certainly be incapable of producing an hourglass shaped PN. One may thus conclude that truly-isolated single stars are highly unlikely to produce even a vaguely aspherical nebula (unless interacting with the interstellar medium; Wareing et al. 2007, and even in those cases the resulting nebula will not be axisymmetrical), while stars with sub-stellar companions might be capable of producing elliptical PNe similar to IC 418 (Fig. 9.2).

9.1.3   Will the Sun Form a Planetary Nebula?

As discussed earlier, the visibility of a PN is a complex function of the post-AGB evolution of its central star as well as the optical depth of the ejecta. The post-AGB models of Miller Bertolami (2016) indicate that the Sun will reach a temperature high enough to ionise Hydrogen some 5,000 years after leaving the AGB, but will take more than twice that before [O III] emission would be observed (due to its much higher excitation potential). By this time, it is quite possible that the shell may no longer be completely optically thick—probably resulting in a rather faint PN. However, as highlighted by Gesicki et al. (2018) if the Sun leaves the AGB during the Helium burning phase, the post-AGB evolution will be much slower and thus never form a PN. With this in mind, they conclude that the Sun is close to the mass limit by which a single star can produce a PN.

In the above discussion, the presence of the solar system has been ignored. While the effect of the close-in rocky planets on the evolution of the Sun is likely to be negligible, the outer gas giants are massive enough to potentially influence its mass-loss rate and/or mass-loss morphology. This could feasibly speed up its post-AGB evolution or lead to strong enough axisymmetries in the ejecta to ensure that some region of the nebula remains optically thick for long enough. However, given their orbital radii (all greater than >5 AU) they are very unlikely to be engulfed by the Sun, instead expanding to wider orbits and perhaps even becoming unbound (Mustill and Villaver 2012), dramatically reducing any impact they may have on the formation of the solar PN.

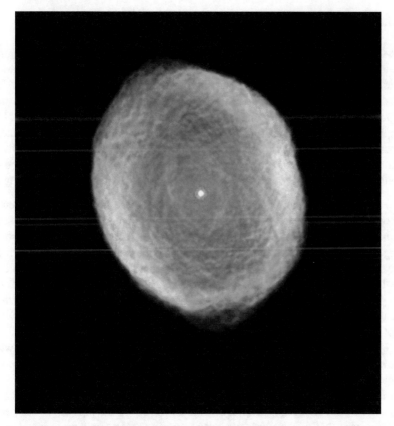

Fig. 9.2 HST image of the elliptical PN IC 418. Image credit: NASA and The Hubble Heritage Team (STScI/AURA); Acknowledgment: Dr. Raghvendra Sahai (JPL) and Dr. Arsen R. Hajian (USNO)

All in all, **the answer to the question "Will the Sun form a planetary nebula?" is thus "maybe"**, but at best it will be rather faint and probably not the most interesting in terms of morphology.

9.2 Outlook

9.2.1 What Can PNe Tell Us About Binary Evolution?

While the focus of this book is on the importance of binarity in understanding PNe, it is particularly interesting to consider the converse: How important are (or could be) PNe in understanding binary evolution?

9.2.1.1 Common Envelope

Post-CE PNe and their central stars represent the direct products of the common envelope (CE)—with the nebulae representing the remnant of the ejected envelope while the central binary has not had time to adjust following the ejection. This makes them truly unique probes of the CE phase—if one can completely constrain the properties of both binary and nebula then, in principle, it should be possible to completely constrain/reconstruct the CE evolution. However, as has been repeatedly demonstrated throughout the previous chapters (e.g., Chap. 2), this is a far from trivial task. All in all, this means that we have thus far been restricted to a few limited but equally intriguing windows into the CE process (described in more detail in Chap. 3):

- Ejecta morphology and kinematics: Spatially-resolved, high-resolution spectroscopy has been used to confirm that the CE is preferentially ejected in the binary orbital plane (Hillwig et al. 2016b). These studies also open the door to studying the ejecta energetics (when combined with mass measurements), which can be compared to predictions based on hydrodynamic models (e.g., Ohlmann et al. 2016; García-Segura et al. 2018). Interestingly, a number of post-CE PNe are found to present with particularly low ionised masses (Corradi et al. 2015a)—perhaps indicating that the final CE event (that resulted in the formation of the PN) was not responsible for removing the entirety of the primary's envelope.
- Pre-CE evolution/mass transfer: Several post-CE PNe display polar outflows which have been demonstrated to be kinematically older than the central nebulae (see Sect. 3.3.1). It is thought that these outflows trace a mass transfer episode prior to the CE event—as such they can be used to probe, for example, the pre-CE magnetic field strength (Tocknell et al. 2014) and even the pre-CE orbital period (Boffin et al. 2012b). Furthermore, the companion of the Necklace is found to display extremely elevated abundances due to chemical contamination by material accreted from the Carbon-enriched primary, thus offering a second route towards probing this pre-CE mass transfer (Miszalski et al. 2013a).
- Mass and period distribution: Albeit with small number statistics (and perhaps biased by the targeted search methods used to discover the majority of post-CE central stars), the period distribution of white dwarf/main-sequence binaries can be used to test CE population synthesis models (as in, for example, Toonen and Nelemans 2013). Similarly, as highlighted in Sect. 3.3.2, one can use the measured masses to try to reconstruct the pre-CE system parameters (Iaconi and De Marco 2019). Again, there are not many systems sufficiently constrained to carry out such an analysis, but in almost every case the pre-CE mass ratio[4] is found to be intriguingly high—perhaps indicative of some physical process by which systems with lower mass ratios avoid (or fail to survive) the CE.
- Double-degenerate central stars: A significant fraction of the known post-CE central stars are thought to be double-degenerate (DD) systems. This is particularly

---

[4]As throughout this book, the mass ratio is defined as the mass of the primary (i.e. the nebular progenitor) over the mass of the companion.

surprising given that such systems are intrinsically more difficult to detect via photometric monitoring (generally only presenting significant photometric variability when one or both components are close to Roche lobe filling, see Sect. 3.1.3.2). As such, this is often taken to indicate that the true DD fraction is even greater (Hillwig et al. 2010; Jones and Boffin 2017a). Given that of the three DD central stars subjected to detailed modelling, two are found to be strong candidate type Ia supernova progenitors, this has important implications for evaluating the likelihood of the various formation scenarios that have been proposed (Santander-García et al. 2015; Jones and Boffin 2017a).

9.2.1.2 Wide Binaries and Wind Roche Lobe Overflow

Just as in post-CE PNe, the PNe surrounding wider binaries give us a unique opportunity to study the mass-loss in these systems. As highlighted in Chap. 5, it is clear that even rather wide binaries can experience significant mass transfer—likely via wind Roche lobe overflow. PNe surrounding such systems allow us to probe both the accretion (through the study of the central stars) and the overall mass-loss (via study of the nebulae). Once again, just as for post-CE PNe, these systems are "fresh" from the most recent episode of mass loss/transfer that culminated in the formation of the PN—meaning that the stars themselves have not been able to adjust, placing stronger constraints on the amount of material and angular momentum accreted. Similarly, the observed stellar parameters (principally abundances and masses) can be used to directly confront AGB nucleosynthesis models (Löbling et al. 2019).

9.2.2 The Missing Pieces of the Puzzle

It is clear that our understanding of PNe in a binary evolution context is still far from complete, with many, important open questions. Here, we try to summarise some of the most critical avenues of future study, which have the potential to significantly advance the study of binary PNe.

9.2.2.1 Further Constraining the Binary Fraction

As highlighted in Chap. 6, one of the key pieces of information in judging the importance of binarity in the formation and evolution of PNe is just how many PNe are derived from a binary pathway. While there have been significant advances in recent years, with a robust determination of the photometrically-variable, surviving, close-binary fraction from the OGLE survey (Miszalski et al. 2009a) as well as refined determinations of the total binary fraction via the search for photometric, infrared excesses (De Marco et al. 2013; Douchin et al. 2015), the *true* binary fraction is still rather uncertain. Photometric searches, like the OGLE study of Miszalski et al.

(2009a), are limited to systems which display appreciable variability, limiting detections to the shortest period, highest inclination binaries and biased against evolved secondaries. The photometric, infrared-excess technique is sensitive to any orbital period, but is instead biased towards low-surface-brightness nebulae where sufficient precision in the photometry can be obtained across all bands (Douchin et al. 2015). Similarly, the technique is fraught with difficulties when measurements are obtained in a limited number of bands (for example, when exploiting a wide-field survey) as correction for extinction can, in some cases, mask the detection of certain spectral types of secondary (Barker et al. 2018). An intriguing alternative to circumvent these difficulties would be to search for infrared excesses spectroscopically. While one would still need to carefully account for nebular contamination, the fact that the emission lines would be spectroscopically resolved from the underlying stellar continuum would mean that even bright, irregular and/or compact nebulae would be accessible. The major disadvantage is, obviously, the vastly increased observing time investment required as well as the need for larger aperture telescopes. In spite of this, modern instrumentation capable of spectroscopically observing large wavelength ranges, like VLT-XSHOOTER which is capable of simultaneously observing from the UVB (at 300 nm) through to the NIR (at 2480 nm) at medium- to high-resolution ($3000 < R < 20000$), means this methodology is, at least, feasible.

9.2.2.2   Towards Statistical Significance

In terms of exploiting binary central stars in understanding not only the formation of PNe, but also binary evolution in general, the most obvious problem is the lack of statistical significance. To date, most of the key findings that have been highlighted throughout this book are based on just a handful of systems which have been studied in detail. Fortunately, there is an obvious (albeit laborious) way to correct this deficiency—find and characterise as many binary central stars (both post-CE and wide) as possible, as well as studying in detail the properties (morphology, kinematics, masses, chemistry, etc.) of their host nebulae. As the work presented in this book highlights, these studies are on-going and there is hope that we may soon be in a position to draw further statistically significant conclusions, relating the binary parameters to those of the surrounding PNe and using this to constrain the binary evolutionary processes which resulted in their formation.

9.2.2.3   What Is the Origin of the Extreme Abundance Discrepancies in Close-Binary PNe?

As detailed in Sect. 5.2, some PNe with close-binary central stars display abundance patterns which are discrepant by factors of up to 100 depending on whether they are derived using ORL or CEL fluxes (Wesson et al. 2018). Such abundance discrepancies are usually of the order 2–3 in field PNe and HII regions, strongly indicating that there may be multiple factors contributing to the observed differences, at least one of which is driven or amplified by the presence of the central binary (causing the much higher discrepancies found in close-binary PNe). The recent discovery that the ORL and CEL emitting gas phases do not appear coincident in these extreme cases (with the lower temperature, Hydrogen-deficient, ORL-emitting phase appearing more centrally concentrated; García-Rojas et al. 2016) has been taken to indicate that two may arise from two non-coeval ejection episodes (Jones et al. 2016). This interpretation is consistent with the observed abundance patterns in the central-most ORL emitting gas, which are reminiscent of Neon novae abundances (Wesson et al. 2008). Kinematical observations could hold the key to constraining the origins of this second gas phase, certainly having the potential to probe the likelihood that it arises from a nova-like eruption from the central pre-WD immediately following the ejection of the CE. Studies comparing ORL and CEL kinematics have already revealed tantalising differences, indicative that the two gas phases do not share the same origin (Richer et al. 2013, 2017), however to date no detailed study has been performed of the kinematics of ORLs and CELs originating from the same chemical species (the most obvious of which would be O^{2+}) in an extreme ADF PN.

9.2.2.4 Is There Really a Missing Mass Problem?

One of the most puzzling results of recent years is the discovery that the ionised masses of post-CE PNe appear to be several orders of magnitude lower than those of the general PN population (Corradi et al. 2015a). This is at odds with expectation given that such post-CE PNe are thought to represent the complete ejection of the primary's envelope and thus should be at least as massive as field PNe (if not more so) which form from much slower AGB wind mass-loss (Santander-García et al. 2019). While there are significant uncertainties in these mass determinations[5] (many of which are associated with the presence of a multiphase gas with differing temperatures and densities which also gives rise to the abundance discrepancy problem previously highlighted), they do not appear to be reconcilable with our current understanding. However, with the ionised masses of only a handful of post-CE PNe known, it is difficult to draw strong conclusions. It is critical to measure the total masses of a significant number of PNe—necessitating multiwavelength studies to derive not only the ionised masses (including the mass in any low-luminosity haloes) but also molecular and neutral masses—as well as attempting to trace the (radial) mass distribution, down to the lowest densities possible, in order to constrain the mass-loss rate and chronology. Collectively, these data can then be used to confront hydrodynamic CE models, in particular those which favour multiple ejection episodes (for example, grazing envelope evolution; Soker 2015) which could perhaps offer the clearest route towards explaining the low, observed ionised masses.

---

[5]As highlighted in Sect. 3.3.2, there is a general issue with deriving total ionised masses based simply on integrated fluxes which tend to "miss" the extended, low-density, low-luminosity haloes that are expected to contain a significant amount of the total ionised nebular mass (Buckley and Schneider 1995; Villaver et al. 2002).

References

Abt, H. A., & Levy, S. G. (1976). *ApJS, 30*, 273.
Adam, C., & Mugrauer, M. (2014). *MNRAS, 444*, 3459.
Afšar, M., & Bond, H. E. (2005). *MmSAI, 76*, 608.
Afşar, M., & Ibanoğlu, C. (2008). *MNRAS, 391*, 802.
Alcock, C., Allsman, R. A., Alves, D. R., et al. (1999). *PASP, 111*, 1539.
Aller, A., Montesinos, B., Miranda, L. F., Solano, E., & Ulla, A. (2015). *MNRAS, 448*, 2822.
Aller, A., Lillo-Box, J., Vučković, M., et al. (2018). *MNRAS, 476*, 1140.
Arnaboldi, M., Freeman, K. C., Mendez, R. H., et al. (1996). *ApJ, 472*, 145.
Baade, W. (1955). *AJ, 60*, 151.
Balick, B., & Frank, A. (2002). *ARA&A, 40*, 439.
Barker, H., Zijlstra, A., De Marco, O., et al. (2018). *MNRAS, 475*, 4504.
Bautista, M. A., & Ahmed, E. E. (2018). *ApJ, 866*, 43.
Bear, E., & Soker, N. (2017). *ApJL, 837*, L10.
Bell, G. O., & Goldreich, P. (1966). *PASP, 78*, 232.
Bell, S. A., Pollacco, D. L., & Hilditch, R. W. (1994). *MNRAS, 270*, 449.
Benson, R. S. (1970). PhD thesis, University of California, Berkeley.
Bessell, M. S. (1990). *PASP, 102*, 1181.
Bhattacharya, S., Arnaboldi, M., Hartke, J., et al. (2019). arXiv e-prints.
Blackman, E. G., & Lucchini, S. (2014). *MNRAS, 440*, L16.
Blagorodnova, N., Kotak, R., Polshaw, J., et al. (2017). *ApJ, 834*, 107.
Blind, N., Boffin, H. M. J., Berger, J.-P., et al. (2011). *A&A, 536*, A55.
Boffin, H. (2015). In P. Dufour, P. Bergeron, & G. Fontaine (Eds.), 19th European Workshop on White Dwarfs (Vol. 493, p. 527). ASPC Series.
Boffin, H. (2019). In G. Beccari, & H. M. J. Boffin (Eds.), The impact of binary stars on stellar evolution. (pp. 1–11).
Boffin, H. M. J. (2010). *A&A, 524*, A14.
Boffin, H. M. J. (2014). In Ecology of blue straggler stars.
Boffin, H. M. J., & Anzer, U. (1994). *A&A, 284*, 1026.
Boffin, H. M. J., & Jorissen, A. (1988). *A&A, 205*, 155.
Boffin, H. M. J., & Zacs, L. (1994). *A&A, 291*, 811.

Boffin, H. M. J. & Pourbaix, D. (2018). In A. Recio-Blanco, P. de Laverny, A. G. A. Brown, & T. Prusti (Eds.), Astrometry and astrophysics in the gaia sky (Vol. 330, pp. 339–340). IAU Symposium.

Boffin, H. M. J., Cerf, N., & Paulus, G. (1993). *A&A, 271*, 125.

Boffin, H. M. J., Miszalski, B., & Jones, D. (2012a). *A&A, 545*, A146.

Boffin, H. M. J., Miszalski, B., Rauch, T., et al. (2012b). *Science, 338*, 773.

Boffin, H. M. J., Jones, D., Wesson, R., et al. (2018). *A&A, 619*, A84.

Bollen, D., Van Winckel, H., & Kamath, D. (2017). *A&A, 607*, A60.

Bond, H. E. (1976). *PASP, 88*, 192.

Bond, H. E. (2000). In J. H. Kastner, N. Soker, & S. Rappaport (Eds.), Asymmetrical planetary nebulae II: From origins to microstructures (Vol. 199, p. 115). ASPC Series.

Bond, H. E. (2015). *AJ, 149*, 132.

Bond, H. E., & Ciardullo, R. (2018). *Research Notes of the American Astronomical Society, 2*, 143.

Bond, H. E., Ciardullo, R., & Meakes, M. G. (1993). In R. Weinberger & A. Acker (Eds.), Planetary nebulae (Vol. 155, p. 397). IAU Symposium.

Bond, H. E., & Livio, M. (1990). *ApJ, 335*, 568.

Bond, H. E., Pollacco, D. L., & Webbink, R. F. (2003). *AJ, 125*, 260.

Borucki, W. J., Koch, D., Basri, G., et al. (2010). *Science, 327*, 977.

Bowen, I. S. (1927). *Nature, 120*, 473.

Bowen, I. S. (1928). *ApJ, 67*, 1.

Brown, A. J., Jones, D., Boffin, H. M. J., & Van Winckel, H. (2019). *MNRAS, 482*, 4951.

Bruch, A., Vaz, L. P. R., & Diaz, M. P. (2001). *A&A, 377*, 898.

Brunner, M., Mecina, M., Maercker, M., et al. (2019). *A&A, 621*, A50.

Buckley, D., & Schneider, S. E. (1995). *ApJ, 446*, 279.

Budaj, J. (2011). *AJ, 141*, 59.

Bujarrabal, V., Castro-Carrizo, A., Alcolea, J., et al. (2013). *A&A, 557*, L11.

Bujarrabal, V., Castro-Carrizo, A., Alcolea, J., & Van Winckel, H. (2015). *A&A, 575*, L7.

Bujarrabal, V., Castro-Carrizo, A., Alcolea, J., et al. (2016). *A&A, 593*, A92.

Bujarrabal, V., Castro-Carrizo, A., Alcolea, J., et al. (2017). *A&A, 597*, L5.

Bujarrabal, V., Castro-Carrizo, A., Van Winckel, H., et al. (2018). *A&A, 614*, A58.

Chamandy, L., Frank, A., Blackman, E. G., et al. (2018). *MNRAS, 480*, 1898.

Chamandy, L., Tu, Y., Blackman, E. G., et al. (2019). *MNRAS, 486*, 1070.

Ciardullo, R. (2013). In R. de Grijs (Ed.), Advancing the physics of cosmic distances (Vol. 289, pp. 247–254). IAU Symposium.

Ciardullo, R. (2012). *Ap&SS, 341*, 151.

Ciardullo, R., & Jacoby, G. H. (1992). *ApJ, 388*, 268.

Ciardullo, R., Jacoby, G. H., Ford, H. C., & Neill, J. D. (1989). *ApJ, 339*, 53.

Ciardullo, R., Bond, H. E., Sipior, M. S., et al. (1999). *AJ, 118*, 488.

Claret, A. (2001). *MNRAS, 327*, 989.

Clayton, G. C., Sugerman, B. E. K., Stanford, S. A., et al. (2011). *ApJ, 743*, 44.

Cliffe, J. A., Frank, A., Livio, M., & Jones, T. W. (1995). *ApJL, 447*, L49.

Coccato, L., Gerhard, O., Arnaboldi, M., et al. (2009). *MNRAS, 394*, 1249.

Corradi, R. L. M., Balick, B., & Santander-García, M. (2011a). *A&A, 529*, A43.

Corradi, R. L. M., Sabin, L., Miszalski, B., et al. (2011b). *MNRAS, 410*, 1349.

Corradi, R. L. M., Rodríguez-Gil, P., Jones, D., et al. (2014). *MNRAS, 441*, 2799.

Corradi, R. L. M., García-Rojas, J., Jones, D., & Rodríguez-Gil, P. (2015a). *ApJ, 803*, 99.

Corradi, R. L. M., Kwitter, K. B., Balick, B., Henry, R. B. C., & Hensley, K. (2015b). *ApJ, 807*, 181.

Cummings, J. D., Kalirai, J. S., Tremblay, P. E., Ramirez-Ruiz, E., & Choi, J. (2018). *ApJ, 866*, 21.

Davis, P. J., Kolb, U., & Willems, B. (2010). *MNRAS, 403*, 179.

Davis, P. J., Kolb, U., & Knigge, C. (2012). *MNRAS, 419*, 287.

Davis, B. D., Ciardullo, R., Jacoby, G. H., Feldmeier, J. J., & Indahl, B. L. (2018). *ApJ, 863*, 189.

De Marco, O. (2009). *PASP, 121*, 316.

De Marco, O., & Izzard, R. G. (2017). *PASA, 34,* e001.

De Marco, O., & Soker, N. (2011). *PASP, 123,* 402.

De Marco, O., Bond, H. E., Harmer, D., & Fleming, A. J. (2004). *ApJL, 602,* L93.

De Marco, O., Hillwig, T. C., & Smith, A. J. (2008). *AJ, 136,* 323.

De Marco, O., Passy, J.-C., Frew, D. J., Moe, M., & Jacoby, G. H. (2013). *MNRAS, 428,* 2118.

De Marco, O., Long, J., Jacoby, G. H., et al. (2015). *MNRAS, 448,* 3587.

Delgado-Inglada, G., Morisset, C., & Stasińska, G. (2014). *MNRAS, 440,* 536.

Dermine, T., Jorissen, A., Siess, L., & Frankowski, A. (2009). *A&A, 507,* 891.

Dermine, T., Izzard, R. G., Jorissen, A., & Van Winckel, H. (2013). *A&A, 551,* A50.

Deschamps, R., Siess, L., Davis, P. J., & Jorissen, A. (2013). *A&A, 557,* A40.

Dopita, M. A., Jacoby, G. H., & Vassiliadis, E. (1992). *ApJ, 389,* 27.

Douchin, D., De Marco, O., Frew, D. J., et al. (2015). *MNRAS, 448,* 3132.

Douglas, N. G., Arnaboldi, M., Freeman, K. C., et al. (2002). *PASP, 114,* 1234.

Drew, J. E., Gonzalez-Solares, E., Greimel, R., et al. (2014). *MNRAS, 440,* 2036.

Duquennoy, A., & Mayor, M. (1991). *A&A, 248,* 485.

Edgar, R. G., Nordhaus, J., Blackman, E. G., & Frank, A. (2008). *ApJL, 675,* L101.

Eggleton, P. (2006). Evolutionary processes in binary and multiple stars.

Eggleton, P. P. (1983). *ApJ, 268,* 368.

Eggleton, P. P. (2000). *NewAR, 44,* 111.

El-Badry, K., Rix, H.-W., & Weisz, D. R. (2018). *ApJL, 860,* L17.

Eldridge, J. J., & Tout, C. A. (2019). *The structure and evolution of stars.* Singapore: World Scientific Publishing Co.

Ertel, S., Kamath, D., Hillen, M., et al. (2019). *AJ, 157,* 110.

Escorza, A., Karinkuzhi, D., Jorissen, A., et al. (2019). arXiv e-prints.

Exter, K. M., Pollacco, D., & Bell, S. A. (2003). *MNRAS, 341,* 1349.

Exter, K. M., Pollacco, D. L., Maxted, P. F. L., Napiwotzki, R., & Bell, S. A. (2005). *MNRAS, 359,* 315.

Ferguson, D. H., Liebert, J., Haas, S., Napiwotzki, R., & James, T. A. (1999). *ApJ, 518,* 866.

Ford, H. C. & Jenner, D. C. (1978). In Bulletin of the American astronomical society (Vol. 10, p. 665).

Ford, H. C. (1978). In Y. Terzian (Ed.), Planetary nebulae (Vol. 76, pp. 19–33). IAU Symposium.

Ford, H. C., Jacoby, G. H., & Jenner, D. C. (1978). *ApJ, 223,* 94.

Frank, A., Chen, Z., Reichardt, T., et al. (2018). *Galaxies, 6,* 113.

Frew, D. J. & Parker, Q. A. (2007). In Asymmetrical planetary nebulae IV (pp. 475–482).

Fuhrmann, K., Chini, R., Kaderhandt, L., & Chen, Z. (2017). *ApJ, 836,* 139.

García-Díaz, M. T., Clark, D. M., López, J. A., Steffen, W., & Richer, M. G. (2009). *ApJ, 699,* 1633.

García-Rojas, J., Corradi, R. L. M., Monteiro, H., et al. (2016). *ApJL, 824,* L27.

García-Rojas, J., & Esteban, C. (2007). *ApJ, 670,* 457.

García-Rojas, J., Wesson, R., Boffin, H. M. J., et al. (2019). arXiv e-prints.

García-Segura, G., Ricker, P. M., & Taam, R. E. (2018). *ApJ, 860,* 19.

García-Segura, G., Villaver, E., Langer, N., Yoon, S.-C., & Manchado, A. (2014). *ApJ, 783,* 74.

Gehrz, R. D., & Woolf, N. J. (1971). *ApJ, 165,* 285.

Gesicki, K., Zijlstra, A. A., & Morisset, C. (2016). *A&A, 585,* A69.

Gesicki, K., Zijlstra, A. A., & Miller Bertolami, M. M. (2018). *Nature Astronomy, 2,* 580.

Gorlova, N., Van Winckel, H., Gielen, C., et al. (2012). *A&A, 542,* A27.

Grauer, A. D., & Bond, H. E. (1983). *ApJ, 271,* 259.

Guerrero, M. A. & Miranda, L. F. (2012). A&A.

Guzman-Ramirez, L., Lagadec, E., Jones, D., Zijlstra, A. A., & Gesicki, K. (2014). *MNRAS, 441,* 364.

Hajduk, M., Zijlstra, A. A., & Gesicki, K. (2010). *MNRAS, 406,* 626.

Halbwachs, J. L., Mayor, M., Udry, S., & Arenou, F. (2003). *A&A, 397,* 159.

Hall, P. D., Tout, C. A., Izzard, R. G., & Keller, D. (2013). *MNRAS, 435,* 2048.

Hamers, A. S., & Dosopoulou, F. (2019). *ApJ, 872,* 119.

Han, Z., Podsiadlowski, P., & Eggleton, P. P. (1995). *MNRAS, 272*, 800.
Hardy, A., Schreiber, M. R., Parsons, S. G., et al. (2016). *MNRAS, 459*, 4518.
Harm, R., & Schwarzschild, M. (1975). *ApJ, 200*, 324.
Henize, K. G., & Westerlund, B. E. (1963). *ApJ, 137*, 747.
Herschel, W. (1791). Philosophical Transactions of the Royal Society of London Series I, 81, 71.
Hilditch, R. W., Harries, T. J., & Hill, G. (1996). *MNRAS, 279*, 1380.
Hillen, M., Kluska, J., Le Bouquin, J.-B., et al. (2016). *A&A, 588*, L1.
Hillwig, T. C. (2011). In A. A. Zijlstra, F. Lykou, I. McDonald, & E. Lagadec (Eds.), Asymmetric Planetary Nebulae 5 Conference, Held in Bowness-on-Windermere, U.K., 20–25 June 2010. Jodrell Bank Centre for Astrophysics.
Hillwig, T. C., Bond, H. E., Afşar, M., & De Marco, O. (2010). *AJ, 140*, 319.
Hillwig, T. C., Frew, D. J., Louie, M., et al. (2015). *AJ, 150*, 30.
Hillwig, T. C., Bond, H. E., Frew, D. J., Schaub, S. C., & Bodman, E. H. L. (2016a). *AJ, 152*, 34.
Hillwig, T. C., Jones, D., De Marco, O., et al. (2016b). *ApJ, 832*, 125.
Hillwig, T. C., Frew, D. J., Reindl, N., et al. (2017). *AJ, 153*, 24.
Hjellming, M. S., & Webbink, R. F. (1987). *ApJ, 318*, 794.
Hodge, P. W. (1966). The physics and astronomy of galaxies and cosmology.
Homan, W., Richards, A., Decin, L., de Koter, A., & Kervella, P. (2018). *A&A, 616*, A34.
Horvat, M., Conroy, K. E., Pablo, H., et al. (2018). *ApJS, 237*, 26.
Horvat, M., Conroy, K. E., Jones, D., & Prša, A. (2019). *ApJS, 240*, 36.
Howell, S. B., Sobeck, C., Haas, M., et al. (2014). *PASP, 126*, 398.
Hrivnak, B. J., Van de Steene, G., Van Winckel, H., et al. (2017). *ApJ, 846*, 96.
Huang, C., Wu, Y., & Triaud, A. H. M. J. (2016). *ApJ, 825*, 98.
Huarte-Espinosa, M., Frank, A., Balick, B., et al. (2012). *MNRAS, 424*, 2055.
Huckvale, L., Prouse, B., Jones, D., et al. (2013). *MNRAS, 434*, 1505.
Huggins, W., & Miller, W. A. (1864). *Philosophical Transactions of the Royal Society of London, 154*, 437.
Hui, X., Ford, H. C., Freeman, K. C., & Dopita, M. A. (1995). *ApJ, 449*, 592.
Hurley, J. R., Pols, O. R., & Tout, C. A. (2000). *MNRAS, 315*, 543.
Hurley, J. R., Tout, C. A., & Pols, O. R. (2002). *MNRAS, 329*, 897.
Iaconi, R., & De Marco, O. (2019). arXiv e-prints arXiv:1902.02039.
Iaconi, R., De Marco, O., Passy, J.-C., & Staff, J. (2018). *MNRAS, 477*, 2349.
Iben, I, Jr., & Livio, M. (1993). *PASP, 105*, 1373.
Iben, I, Jr., & Tutukov, A. V. (1993). *ApJ, 418*, 343.
Iłkiewicz, K., Mikołajewska, J., Miszalski, B., Kozłowski, S., & Udalski, A. (2018). *MNRAS, 476*, 2605.
Ivanova, N. (2015). In H. M: J. Boffin, G. Carraro, & G. Beccari (Eds.), Binary evolution: Roche lobe overflow and blue stragglers (p. 179).
Ivanova, N. (2017). In J. J. Eldridge, J. C. Bray, L. A. S. McClelland, & L. Xiao (Eds.), The lives and death-throes of massive stars (Vol. 329, pp. 199–206). IAU Symposium.
Ivanova, N. (2018). *ApJL, 858*, L24.
Ivanova, N., & Nandez, J. (2018). *Galaxies, 6*, 75.
Ivanova, N., Justham, S., Chen, X., et al. (2013). *A&ARv, 21*, 59.
Ivezić, Ž., Kahn, S. M., Tyson, J. A., et al. (2008). arXiv e-prints.
Jackson, B., Arras, P., Penev, K., Peacock, S., & Marchant, P. (2017). *ApJ, 835*, 145.
Jacoby, G. H. (1979). In Bulletin of the American astronomical society (Vol. 11, p. 634). BAAS.
Jacoby, G. H., Hillwig, T., De Marco, O., et al. (2018). In American astronomical society meeting abstracts #231, 241.04 (Vol. 231). American Astronomical Society Meeting Abstracts.
Jacoby, G. H. (1980). *ApJS, 42*, 1.
Jacoby, G. H. (1989). *ApJ, 339*, 39.
Jacoby, G. H., & Lesser, M. P. (1981). *AJ, 86*, 185.
Jacoby, G. H., Branch, D., Ciardullo, R., et al. (1992). *PASP, 104*, 599.
Jacoby, G. H., Ciardullo, R., De Marco, O., et al. (2013). *ApJ, 769*, 10.

Jahanara, B., Mitsumoto, M., Oka, K., et al. (2005). *A&A, 441*, 589.

Jones, D. (2019). The importance of binarity in the formation and evolution of planetary nebulae. Cambridge Astrophysics (pp. 106–127). Cambridge: Cambridge University Press.

Jones, D., Pejcha, O., & Corradi, R. L. M. (2019). MNRAS, in press (arxiv:1908.04582).

Jones, D., & Boffin, H. M. J. (2017a). *Nature Astronomy, 1*, 0117.

Jones, D., & Boffin, H. M. J. (2017b). *MNRAS, 466*, 2034.

Jones, D., Boffin, H. M. J., Miszalski, B., et al. (2014). *A&A, 562*, A89.

Jones, D., Boffin, H. M. J., Rodríguez-Gil, P., et al. (2015). *A&A, 580*, A19.

Jones, D., Wesson, R., García-Rojas, J., Corradi, R. L. M., & Boffin, H. M. J. (2016). *MNRAS, 455*, 3263.

Jones, D., Van Winckel, H., Aller, A., Exter, K., & De Marco, O. (2017). *A&A, 600*, L9.

Jones, D., Boffin, H. M. J., Sowicka, P., et al. (2019). *MNRAS, 482*, L75.

Jorissen, A., Boffin, H. M. J., Karinkuzhi, D., et al. (2019). *A&A, 626*, 127.

Juan de Dios, L. & Rodríguez, M. (2017). MNRAS, 469, 1036.

Kahn, F. D., & West, K. A. (1985). *MNRAS, 212*, 837.

Kamath, D., Wood, P. R., & Van Winckel, H. (2014). *MNRAS, 439*, 2211.

Kamath, D., Wood, P. R., & Van Winckel, H. (2015). *MNRAS, 454*, 1468.

Kamath, D., Wood, P. R., Van Winckel, H., & Nie, J. D. (2016). *A&A, 586*, L5.

Kashi, A., & Soker, N. (2011). *MNRAS, 417*, 1466.

Kashi, A., & Soker, N. (2018). *MNRAS, 480*, 3195.

Kingsburgh, R. L., & Barlow, M. J. (1994). *MNRAS, 271*, 257.

Kippenhahn, R., & Meyer-Hofmeister, E. (1977). *A&A, 54*, 539.

Koning, N., Kwok, S., & Steffen, W. (2013). *ApJ, 765*, 92.

Kopal, Z. (1959). Close Binary Systems.

Kukarkin, B. V., Kholopov, P. N., Pskovsky, Y. P., et al. (1971). In General catalogue of variable stars (3rd ed.).

Kutter, G. S., & Sparks, W. M. (1974). *ApJ, 192*, 447.

Kwok, S. (1993). *ARA&A, 31*, 63.

Kwok, S., Purton, C. R., & Fitzgerald, P. M. (1978). *ApJ, 219*, 125.

Kwok, S., Hrivnak, B. J., & Su, K. Y. L. (2000). *ApJL, 544*, L149.

Lagadec, E. (2018). *Galaxies, 6*, 99.

Lagadec, E., & Zijlstra, A. A. (2008). *MNRAS, 390*, L59.

Lagadec, E., Verhoelst, T., Mékarnia, D., et al. (2011). *MNRAS, 417*, 32.

Lajoie, C.-P., & Sills, A. (2011). *ApJ, 726*, 67.

Lauterborn, D. (1970). *A&A, 7*, 150.

Lin, D. N. C. (1977). *MNRAS, 179*, 265.

Liu, X.-W. (2006). In M. J. Barlow & R. H. Méndez (Ed.), Planetary nebulae in our galaxy and beyond (Vol. 234, pp. 219–226). IAU Symposium.

Liu, X.-W., & Danziger, J. (1993). *MNRAS, 263*, 256.

Liu, X.-W., Storey, P. J., Barlow, M. J., et al. (2000). *MNRAS, 312*, 585.

Liu, X.-W., Barlow, M. J., Zhang, Y., Bastin, R. J., & Storey, P. J. (2006). *MNRAS, 368*, 1959.

Livio, M. (1989). *SSRv, 50*, 299.

Livio, M., & Soker, N. (1988). *ApJ, 329*, 764.

Löbling, L., Boffin, H. M. J., & Jones, D. (2019). *A&A, 624*, A1.

Longobardi, A., Arnaboldi, M., Gerhard, O., et al. (2013). *A&A, 558*, A42.

Longobardi, A., Arnaboldi, M., Gerhard, O., Pulsoni, C., & Söldner-Rembold, I. (2018). *A&A, 620*, A111.

Lopez, J. A., Meaburn, J., & Palmer, J. W. (1993). ApJL, 415, L135+.

Lucy, L. B. (1967). *AJ, 72*, 813.

Lutz, J., Alves, D., Becker, A., et al. (1998). In Bulletin of the American astronomical society. American Astronomical Society Meeting Abstracts #192 (Vol. 30, p. 894).

Lutz, J., Fraser, O., McKeever, J., & Tugaga, D. (2010). *PASP, 122*, 524.

MacLeod, M., Ostriker, E. C., & Stone, J. M. (2018). *ApJ, 868*, 136.

Maercker, M., Mohamed, S., Vlemmings, W. H. T., et al. (2012). *Nature*, *490*, 232.

Magrini, L., Stanghellini, L., & Villaver, E. (2009). *ApJ*, *696*, 729.

Manick, R., Miszalski, B., & McBride, V. (2015). *MNRAS*, *448*, 1789.

Maoz, D., Mannucci, F., & Nelemans, G. (2014). *ARA&A*, *52*, 107.

Margon, B., Kupfer, T., Burdge, K., et al. (2018). *ApJL*, *856*, L2.

McClure, R. D., & Woodsworth, A. W. (1990). *ApJ*, *352*, 709.

McCullough, P. R., Bender, C., Gaustad, J. E., Rosing, W., & Van Buren, D. (2001). *AJ*, *121*, 1578.

McDonald, I., De Beck, E., Zijlstra, A. A., & Lagadec, E. (2018). *MNRAS*, *481*, 4984.

Meatheringham, S. J., Dopita, M. A., Ford, H. C., & Webster, B. L. (1988). *ApJ*, *327*, 651.

Méndez, R. H., & Niemela, V. S. (1981). *ApJ*, *250*, 240.

Merle, T., Jorissen, A., Masseron, T., et al. (2014). *A&A*, *567*, A30.

Messier, C. (1781). Catalogue des Nébuleuses et des Amas d'Étoiles (Catalog of Nebulae and Star Clusters), Technical report.

Metzger, B. D., & Pejcha, O. (2017). *MNRAS*, *471*, 3200.

Middlemass, D., Clegg, R. E. S., & Walsh, J. R. (1989). *MNRAS*, *239*, 1.

Mikołajewska, J. (2003). In R. L. M. Corradi, J. Mikolajewska, & T. J. Mahoney (Eds.), Symbiotic stars probing stellar evolution (Vol. 303, p. 9). ASPC Series.

Mikołajewska, J. (2012). *Baltic Astronomy*, *21*, 5.

Miller Bertolami, M. M. (2017). In X. Liu, L. Stanghellini, & A. Karakas (Ed.), Planetary nebulae: Multi-wavelength probes of stellar and galactic evolution (Vol. 323, pp. 179–183). IAU Symposium.

Miller Bertolami, M. M. (2016). *A&A*, *588*, A25.

Miszalski, B., Acker, A., Moffat, A. F. J., Parker, Q. A., & Udalski, A. (2009a). *A&A*, *496*, 813.

Miszalski, B., Acker, A., Parker, Q. A., & Moffat, A. F. J. (2009b). *A&A*, *505*, 249.

Miszalski, B., Corradi, R. L. M., Boffin, H. M. J., et al. (2011a). *MNRAS*, *413*, 1264.

Miszalski, B., Corradi, R. L. M., Jones, D., et al. (2011b). In A. A. Zijlstra, F. Lykou, I. McDonald, & E. Lagadec, (Eds.), Asymmetric Planetary Nebulae 5 Conference, Held in Bowness-on-Windermere, U.K., 20–25 June 2010. Jodrell Bank Centre for Astrophysics.

Miszalski, B., Jones, D., Rodríguez-Gil, P., et al. (2011c). *A&A*, *531*, A158.

Miszalski, B., Boffin, H. M. J., Frew, D. J., et al. (2012). *MNRAS*, *419*, 39.

Miszalski, B., Boffin, H. M. J., & Corradi, R. L. M. (2013a). *MNRAS*, *428*, L39.

Miszalski, B., Boffin, H. M. J., Jones, D., et al. (2013b). *MNRAS*, *436*, 3068.

Miszalski, B., Manick, R., Mikołajewska, J., et al. (2018a). *MNRAS*, *473*, 2275.

Miszalski, B., Manick, R., Mikołajewska, J., Van Winckel, H., & Iłkiewicz, K. (2018b). *PASA*, *35*, e027.

Miszalski, B., Manick, R., Van Winckel, H., & Escorza, A. (2019a). *PASA*, *36*, e018.

Miszalski, B., Manick, R., Van Winckel, H., & Mikołajewska, J. (2019b). *MNRAS*, *487*, 1040.

Mitchell, D. L., Pollacco, D., O'Brien, T. J., et al. (2007). *MNRAS*, *374*, 1404.

Močnik, T., Lloyd, M., Pollacco, D., & Street, R. A. (2015). *MNRAS*, *451*, 870.

Moe, M., & De Marco, O. (2006). *ApJ*, *650*, 916.

Moe, M., & Di Stefano, R. (2017). *ApJS*, *230*, 15.

Morton, D. C. (1960). *ApJ*, *132*, 146.

Murphy, S. J., Moe, M., Kurtz, D. W., et al. (2018). *MNRAS*, *474*, 4322.

Mustill, A. J., & Villaver, E. (2012). *ApJ*, *761*, 121.

Nagae, T., Oka, K., Matsuda, T., et al. (2004). *A&A*, *419*, 335.

Nandez, J. L. A., & Ivanova, N. (2016). *MNRAS*, *460*, 3992.

Nandez, J. L. A., Ivanova, N., & Lombardi, J. C. (2015). *MNRAS*, *450*, L39.

Napiwotzki, R., Heber, U., & Koeppen, J. (1994). *A&A*, *292*, 239.

Nebot Gómez-Morán, A., Gänsicke, B. T., Schreiber, M. R., et al. (2011). *A&A*, *536*, A43.

Nelemans, G. (2018). arXiv e-prints.

Nelemans, G., & Tout, C. A. (2005). *MNRAS*, *356*, 753.

Neo, S., Miyaji, S., Nomoto, K., & Sugimoto, D. (1977). *PASJ*, *29*, 249.

Nicholson, J. W. (1911). *MNRAS*, *72*, 49.

Nie, J. D., Wood, P. R., & Nicholls, C. P. (2012). *MNRAS*, *423*, 2764.

Nordhaus, J., & Blackman, E. G. (2006). *MNRAS*, *370*, 2004.

Nordhaus, J., Blackman, E. G., & Frank, A. (2007). *MNRAS*, *376*, 599.

Ohlmann, S. T., Röpke, F. K., Pakmor, R., & Springel, V. (2016). *ApJL*, *816*, L9.

Oomen, G.-M., Van Winckel, H., Pols, O., et al. (2018). *A&A*, *620*, A85.

Paczyński, B. (1971). *ARA&A*, *9*, 183.

Paczynski, B. (1976). In P. Eggleton, S. Mitton, & J. Whelan (Eds.), Structure and evolution of close binary systems (Vol. 73, p. 75). IAU Symposium.

Paczyński, B., & Sienkiewicz, R. (1972). *Acta Astronomica*, *22*, 73.

Parker, Q. A., Acker, A., Frew, D. J., et al. (2006). *MNRAS*, *373*, 79.

Parsons, S. G., Rebassa-Mansergas, A., Schreiber, M. R., et al. (2016). *MNRAS*, *463*, 2125.

Passy, J.-C., De Marco, O., Fryer, C. L., et al. (2012a). *ApJ*, *744*, 52.

Passy, J.-C., Herwig, F., & Paxton, B. (2012b). *ApJ*, *760*, 90.

Pastetter, L., & Ritter, H. (1989). *A&A*, *214*, 186.

Pavlovskii, K., & Ivanova, N. (2015). *MNRAS*, *449*, 4415.

Paxton, B., Bildsten, L., Dotter, A., et al. (2011). *ApJS*, *192*, 3.

Peimbert, M. & Torres-Peimbert, S. (1983). In D. R. Flower (Ed.), Planetary nebulae (Vol. 103, pp. 233–241). IAU Symposium.

Peimbert, M. (1967). *ApJ*, *150*, 825.

Peimbert, M. (1971). *Boletin de los Observatorios Tonantzintla y Tacubaya*, *6*, 29.

Peimbert, M., Peimbert, A., & Delgado-Inglada, G. (2017). *PASP*, *129*, 082001.

Pejcha, O., & Thompson, T. A. (2015). *ApJ*, *801*, 90.

Pereira, C. B., Miranda, L. F., Smith, V. V., & Cunha, K. (2008). *A&A*, *477*, 535.

Pereira, C. B., Baella, N. O., Daflon, S., & Miranda, L. F. (2010). *A&A*, *509*, A13.

Perek, L. & Kohoutek, L. (1967). Catalogue of galactic planetary nebulae.

Pollacco, D. L., Skillen, I., Collier Cameron, A., et al. (2006). *PASP*, *118*, 1407.

Pottasch, S. R. (1980). *A&A*, *89*, 336.

Prialnik, D., & Livio, M. (1985). *MNRAS*, *216*, 37.

Prša, A. (2019). Modeling and analysis of eclipsing binary stars: The theory and design principles of PHOEBE. Institute of Physics Publishing.

Prša, A., Conroy, K. E., Horvat, M., et al. (2016). *ApJS*, *227*, 29.

Raga, A. C., Esquivel, A., Velázquez, P. F., et al. (2009). *ApJL*, *707*, L6.

Raghavan, D., McAlister, H. A., Henry, T. J., et al. (2010). *ApJS*, *190*, 1.

Rahimi, A., Kawata, D., Allende Prieto, C., et al. (2011). *MNRAS*, *415*, 1469.

Ramstedt, S., Mohamed, S., Vlemmings, W. H. T., et al. (2017). *A&A*, *605*, A126.

Rebassa-Mansergas, A., Ren, J. J., Parsons, S. G., et al. (2016). *MNRAS*, *458*, 3808.

Reichardt, T. A., De Marco, O., Iaconi, R., Tout, C. A., & Price, D. J. (2019). *MNRAS*, *484*, 631.

Reindl, N., Finch, N., Schaffenroth, V., et al. (2018). *Galaxies*, *6*, 88.

Renedo, I., Althaus, L. G., Miller Bertolami, M. M., et al. (2010). *ApJ*, *717*, 183.

Renzini, A. & Buzzoni, A. (1986). In C. Chiosi & A. Renzini (Eds.), Spectral evolution of galaxies (Vol. 122, pp. 195–231). Astrophysics and Space Science Library.

Ressler, M. E., Cohen, M., Wachter, S., et al. (2010). *AJ*, *140*, 1882.

Richer, M. G., Georgiev, L., Arrieta, A., & Torres-Peimbert, S. (2013). *ApJ*, *773*, 133.

Richer, M. G., Suárez, G., López, J. A., & García Díaz, M. T. (2017). *AJ*, *153*, 140.

Ricker, P. M., & Taam, R. E. (2012). *ApJ*, *746*, 74.

Ritter, H. (1976). *MNRAS*, *175*, 279.

Rodríguez-Gil, P., Santander-García, M., Knigge, C., et al. (2010). *MNRAS*, *407*, L21.

Romanowsky, A. J., Douglas, N. G., Arnaboldi, M., et al. (2003). *Science*, *301*, 1696.

Ruciński, S. M. (1969). *AcA*, *19*, 245.

Sabach, E., & Soker, N. (2018). *MNRAS*, *473*, 286.

Sahai, R., & Trauger, J. T. (1998). *AJ*, *116*, 1357.

Sahai, R., Sánchez Contreras, C., & Morris, M. (2005). *ApJ*, *620*, 948.

Sahai, R., Morris, M., Sánchez Contreras, C., & Claussen, M. (2007). *AJ*, *134*, 2200.

Sánchez Contreras, C., Gil de Paz, A., & Sahai, R. (2004). ApJ, 616, 519.
Santander-García, M., Jones, D., Alcolea, J., Wesson, R., & Bujarrabal, V. (2019). In Proceedings of the XIII Scientific Meeting of the Spanish Astronomical Society held on July 16–20, 2018, in Salamanca, Spain. Highlights on Spanish Astrophysics X. p. 392–396, ISBN 978-84-09-09331-1. B. Montesinos, A. Asensio Ramos, F. Buitrago, R. Schödel, E. Villaver, S. Pérez-Hoyos, I. Ordóñez-Etxeberria (Eds.).
Santander-García, M., Rodríguez-Gil, P., Corradi, R. L. M., et al. (2015). Nature, 519, 63.
Schoenberner, D. (1983). ApJ, 272, 708.
Schönberg, M., & Chandrasekhar, S. (1942). ApJ, 96, 161.
Schönberner, D., Jacob, R., Sandin, C., & Steffen, M. (2010). A&A, 523, A86.
Schreiber, M. R., Gänsicke, B. T., Southworth, J., Schwope, A. D., & Koester, D. (2008). A&A, 484, 441.
Schwarzschild, M., & Härm, R. (1965). ApJ, 142, 855.
Shiber, S., Iaconi, R., De Marco, O., & Soker, N. (2019). arXiv e-prints.
Shimanskii, V. V., Borisov, N. V., Sakhibullin, N. A., & Sheveleva, D. V. (2008). Astronomy Reports, 52, 479.
Shklovsky, I. S. (1956). AZh, 33, 315.
Shporer, A., Kaplan, D. L., Steinfadt, J. D. R., et al. (2010). ApJL, 725, L200.
Siegel, M. H., Hoversten, E., Bond, H. E., Stark, M., & Breeveld, A. A. (2012). AJ, 144, 65.
Smith, V. V., Cunha, K., Jorissen, A., & Boffin, H. M. J. (1996). A&A, 315, 179.
Soberman, G. E., Phinney, E. S., & van den Heuvel, E. P. J. (1997). A&A, 327, 620.
Soker, N. (1996). ApJ, 468, 774.
Soker, N. (1997). ApJS, 112, 487.
Soker, N. (1999). AJ, 118, 2424.
Soker, N. (2015). ApJ, 800, 114.
Soker, N. (2016). MNRAS, 455, 1584.
Soker, N. (2017). MNRAS, 471, 4839.
Soker, N. (2019). MNRAS, 483, 5020.
Soker, N., & Livio, M. (1994). ApJ, 421, 219.
Soker, N., & Subag, E. (2005). AJ, 130, 2717.
Sorensen, P. & Pollacco, D. (2004). In M. Meixner, J. H. Kastner, B. Balick, & N. Soker (Eds.), Asymmetrical planetary nebulae III: Winds, structure and the thunderbird. ASPC series (Vol. 313, p. 515).
Soszynski, I., Udalski, A., Kubiak, M., et al. (2004). AcA, 54, 347.
Soszyński, I., Stępień, K., Pilecki, B., et al. (2015). Acta Astronomica, 65, 39.
Sowicka, P., Jones, D., Corradi, R. L. M., et al. (2017). MNRAS, 471, 3529.
Staff, J. E., De Marco, O., Wood, P., Galaviz, P., & Passy, J.-C. (2016). MNRAS, 458, 832.
Taam, R. E. & Sandquist, E. (1998). In K. L. Chan, K. S. Cheng & H. P. Singh (Eds.), 1997 Pacific Rim Conference on Stellar Astrophysics (Vol. 138, p. 349). Astronomical Society of the Pacific Conference Series.
Taam, R. E. (1994). In A. W. Shafter (Ed.), Interacting binary stars. (p. 208). Astronomical Society of the Pacific Conference Series.
Taam, R. E., & Bodenheimer, P. (1989). ApJ, 337, 849.
Taam, R. E., & Bodenheimer, P. (1991). ApJ, 373, 246.
Tauris, T. M., & Dewi, J. D. M. (2001). A&A, 369, 170.
Theuns, T., Boffin, H. M. J., & Jorissen, A. (1996). MNRAS, 280, 1264.
Tocknell, J., De Marco, O., & Wardle, M. (2014). MNRAS, 439, 2014.
Toonen, S., & Nelemans, G. (2013). A&A, 557, A87.
Toonen, S., Claeys, J. S. W., Mennekens, N., & Ruiter, A. J. (2014). A&A, 562, A14.
Torres-Peimbert, S., Peimbert, M., & Daltabuit, E. (1980). ApJ, 238, 133.
Tout, C. A. (2012). In M. T. Richards & I. Hubeny (Eds.), From interacting binaries to exoplanets: Essential modeling tools (Vol. 282, pp. 417–424). IAU Symposium.
Tout, C. A., & Eggleton, P. P. (1988). MNRAS, 231, 823.

Tovmassian, G., Yungelson, L., Rauch, T., et al. (2010). *ApJ, 714*, 178.

Tsebrenko, D., & Soker, N. (2015). *MNRAS, 447*, 2568.

Tutukov, A. & Yungelson, L. (1979). In P. S. Conti & C. W. H. De Loore (Eds.), Mass Loss and Evolution of O-Type Stars (Vol. 83, pp. 401–406). IAU Symposium.

Tylenda, R., Hajduk, M., Kamiński, T., et al. (2011). *A&A, 528*, A114.

Tyndall, A. A., Jones, D., Boffin, H. M. J., et al. (2013). *MNRAS, 436*, 2082.

Udalski, A., Szymanski, M. K., Soszynski, I., & Poleski, R. (2008). *Acta Astronautica, 58*, 69.

van Aarle, E., Van Winckel, H., De Smedt, K., Kamath, D., & Wood, P. R. (2013). *A&A, 554*, A106.

Van der Swaelmen, M., Boffin, H. M. J., Jorissen, A., & Van Eck, S. (2017). *A&A, 597*, A68.

Van Winckel, H. (2014). In J. Cami & N. L. J. Cox (Eds.), The diffuse interstellar bands (Vol. 297, pp. 180–186). IAU Symposium.

Van Winckel, H. (2018). arXiv:1809.00871.

van Winckel, H. (2003). *ARA&A, 41*, 391.

Van Winckel, H. (2007). *Baltic Astronomy, 16*, 112.

Van Winckel, H., Lloyd Evans, T., Reyniers, M., Deroo, P., & Gielen, C. (2006). *MemSAI, 77*, 943.

Van Winckel, H., Jorissen, A., Gorlova, N., et al. (2010). *MmSAI, 81*, 1022.

Van Winckel, H., Jorissen, A., Exter, K., et al. (2014). *A&A, 563*, L10.

Vassiliadis, E., & Wood, P. R. (1994). *ApJS, 92*, 125.

Vauclair, G. (1972). *A&A, 17*, 437.

Villaver, E., Manchado, A., & García-Segura, G. (2002). *ApJ, 581*, 1204.

Vos, J., Østensen, R. H., Vučković, M., & Van Winckel, H. (2017). *A&A, 605*, A109.

Vos, J., Vučković, M., Chen, X., et al. (2019). *MNRAS, 482*, 4592.

Wareing, C. J., Zijlstra, A. A., & O'Brien, T. J. (2007). *MNRAS, 382*, 1233.

Webbink, R. F. (1975). PhD thesis, University of Cambridge.

Webbink, R. F. (1984). *ApJ, 277*, 355.

Webbink, R. F. (2008). In E. F. Milone, D. A. Leahy, & D. W. Hobill (Eds.), Astrophysics and space science library (Vol. 352, p. 233). Astrophysics and Space Science Library.

Weinberger, R. (1989). *A&AS, 78*, 301.

Wesson, R., Liu, X.-W., & Barlow, M. J. (2003). *MNRAS, 340*, 253.

Wesson, R., Liu, X.-W., & Barlow, M. J. (2005). *MNRAS, 362*, 424.

Wesson, R., Barlow, M. J., Liu, X.-W., et al. (2008). *MNRAS, 383*, 1639.

Wesson, R., Jones, D., García-Rojas, J., Boffin, H. M. J., & Corradi, R. L. M. (2018). *MNRAS, 480*, 4589.

Whitworth, A. P., & Lomax, O. (2015). *MNRAS, 448*, 1761.

Willems, B., & Kolb, U. (2004). *A&A, 419*, 1057.

Wilson, R. E. (1979). *ApJ, 234*, 1054.

Wilson, R. E. (1990). *ApJ, 356*, 613.

Wilson, R. E. (2008). *ApJ, 672*, 575.

Wilson, R. E., & Devinney, E. J. (1971). *ApJ, 166*, 605.

Wilson, R. E., & Van Hamme, W. (2014). *ApJ, 780*, 151.

Woods, T. E., & Ivanova, N. (2011). *ApJL, 739*, L48.

Wyse, A. B. (1942). *ApJ, 95*, 356.

Yungelson, L. R. (1973). *Soviet Astronomy, 16*, 864.

Zanstra, H. (1927). *ApJ, 65*, 50.

Ziegler, M., Rauch, T., Werner, K., Köppen, J., & Kruk, J. W. (2012). *A&A, 548*, A109.

Zucker, S., Mazeh, T., & Alexander, T. (2007). *ApJ, 670*, 1326.

Index

H. M. J. Boffin and D. Jones, *The Importance of Binaries in the Formation
and Evolution of Planetary Nebulae*, SpringerBriefs in Astronomy,
https://doi.org/10.1007/978-3-030-25059-1

Printed in the United States
By Bookmasters